CLIMATE CHANGE

in History

A GEOTHEOLOGICAL PERSPECTIVE

Kendall Hunt
publishing company

BARRY VANN

www.kendallhunt.com
Send all inquiries to:
4050 Westmark Drive
Dubuque, IA 52004-1840

Copyright © 2020 by Kendall Hunt Publishing Company

ISBN 978-1-7924-2200-3

Published in the United States of America

Table of Contents

Preface

Brackets have been added to Chapter 2 for identifying borrowed material from Barry A. Vann's text "*The Sources of Nature: Our Quest to Conquer the Planet.*" See example below:

Example:

⌐The Medieval Warm period which ran from ca. 800 to 1250 is well-documented in Scandinavian migration history, but in the scientific literature its existence has been disputed by some scholars while others dismissed it as a minor regional phenomenon.[1] However, as scientists from the southern hemisphere have joined in the quest to challenge the historicity of the Medieval Warm period, it is clear that it occurred in both the northern and southern hemispheres.⌐

---

[1] Crowley and Lowery, "Medieval Warm Period," 51–54; Hughes and Diaz, "Medieval Warm Period," 109–142.

CHAPTER
1

Introduction

OVERVIEW

Because weather is an important factor in sustaining life, few topics in the history of humanity have received more attention than that of weather and climate. Aberrant weather conditions can cause a great deal of panic and worry among those who experience them. No doubt such is the case today among many people who worry about the real and imagined consequences of climate change. Our connection to weather, especially adequate precipitation, is so profound that it invites us, no matter what our ancestral heritage, to view human-environmental relationships in metaphysical terms. From the biblical flood event to President John F. Kennedy's speech in which he described our connection to the sea, we humans are products of the earth and are, therefore, highly dependent on its natural cycles and systems.

In his speech at the America's Cup Dinner held in Newport, Rhode Island, on September 14, 1962, Kennedy gave this metaphysical observation of the mysterious hold that the sea has on people: "All of us have in our veins the exact same percentage of salt in our blood that exists in the ocean, and, therefore, we have salt in our blood, in our sweat, in our tears. We are tied to the ocean. And when we go back to the sea—whether it is to sail or to watch it—we are going back from whence we came."[1] A little more than a dozen years later, Norman Maclean, who was a professor of English literature at the University of Chicago, penned one of the finest contemporary descriptions of our metaphysical connection to water in his moving novella: *A River Runs Through it and Other Stories*. Impressed with the semi-autobiographical portion of the small book, Robert Redford turned the novella into a 1992 film entitled *A River Runs through It.*[2] Having received most of his early education from his Presbyterian minister father, Maclean wrote this poetic explanation of life and our connection to the natural world: "Eventually all things merge into one, and a river runs through it. The river was cut by the world's great flood and runs over rocks from the basement of time. On some of the rocks are timeless raindrops. Under the rocks are the words, and some of the words are theirs"[3].

---

[1] Kennedy Presidential Library and Museum. *America's Cup Dinner.*

[2] Redford, *River Runs.*

[3] Maclean, *River Runs*, 104.

© Maks Ershov/Shutterstock.com

FIGURE 1.1 Middle Fork Flathead River, Glacier National Park, Montana.

© Dennis W Donohue/Shutterstock.com

FIGURE 1.2 Sail Boats near JFK's home in Cape Cod, Massachusetts.

Of course, poets and philosophers are not alone in describing the human connection to the ebbs and flows of the natural world. There are those who dispassionately write about weather and climate: paleo-biologists, historical geographers, geoscientists, paleo-climatologists, and historians are prominent among the academics and scientists who study weather and climate in a historical context. Taken as a whole that eclectic community benefits from the varied nature of the disciplines involved in climate studies. They are able to make inferences about weather and hence climate patterns by examining plant and animal fossils as well as tree rings. They also study gases, pollen, and volcanic particulates as well as sediments and chemicals in ice cores. Archeologists and historians also rely on historical accounts and observations of long-dead authors to help inform their reconstruction of past climates. Perhaps because religious ideas and writings are seen as falling in the purview of philosophers, theologians, and other religionists, natural scientists are inclined to look for nonreligious texts and accounts to support their findings. However, this academic bifurcation is not necessary. Sociologists and geographers as well as historians have been engaged in objectively analyzing religious thinking, movements, practices, and texts for decades.[4]

Applying Geography

Central to this study is a behavioral form of geography that historical geographer John K. Wright introduced into the geographic lexicon in 1947; it is called geotheology.[5] Wright's conception of geotheology (earth and the study of God) more broadly means the conjoining of attributes of sacredness and worship with spatial aspects of the earth (natural settings, weather, climate, towns, and nations). Other writers have made good use of Wright's lexicon. Among the more recent scholars to apply aspects of geotheology to historical studies are Yi-Fu Tuan, Avihu Zakai, and most recently Barry A. Vann and Dorothy Jean Weaver.[6]

In the pages that follow, we will explore how ancient and more recent history has been affected by short-term and long-term weather events. This is an approach to the study of environmental history that the geographers H. C. Darby, Chris Philo, W. Gordon East, and Barry A. Vann would call "geography behind history."[7] Darby went further than just describing geography behind history; Darby actually proposed three ways of looking at historical geography: "history behind geography," "past geographies," and "geography behind history."[8] Darby's three-pronged description of historical geography is comprehensive. Consider that past geographies, simply put, describes a past environmental setting. For instance, if one were to explore how village life in the Swiss Alps looked during the Little Ice Age, one would be engaged in a study of a "past geographies." Fast forward to 1948 and a small town named Donora, Pennsylvania, and one would witness an environmental crisis that is best described as history behind geography. Donora sits on the banks of the Monongahela River and is surrounded by low hills and ridges. Zinc smelting plants and steel mills produced a great number of particulates and noxious gases. When a temperature inversion formed over the small town on October 27, 1948, air at and just above ground level was trapped. Pollutants in the air made hundreds of people sick while causing the deaths of 20 residents. Taking a deeper view of history, one could argue that "past geographies" and "geography behind history" are more prominent in the fascinating story of humanity. That is because the further back in time one ventures, the less ability people have to affect geography, although microscale environments no doubt were affected by fires that

---

[4] Durkheim, *Elementary Forms*, 201; See also the work written by Park, *Sacred Worlds*.

[5] Wright, "Terrae Incognitae," 1–45.

[6] Tuan, *Morality and Imagination*; Zakai, *Exile and Kingdom*; Vann, *Search of Ulster-Scots Land*; and Weaver, "Inheriting the Earth," 6–29.

[7] Darby, *Relations of History and Geography*. Darby was a leading light of historical geography during the 1960s. His *Relations of History and Geography* is made up of unpublished essays that he wrote during the 1960s; they represent the foundations for his idea. See also Philo, "History, Geography, and the Still Greater Mystery". East, *Geography behind History*; Vann, *The Forces of Nature*, 23.

[8] Darby, *Relations of History and Geography*. Darby was a leading light of historical geography during the 1960s. His *Relations of History and Geography* is made up of unpublished essays that he wrote during the 1960s; they represent the foundations for his ideas.

© Everett Historical/Shutterstock.com

FIGURE 1.3 Donora, Pennsylvania ca. 1910 This picture simultaneously shows "history behind geography" and "past geographies."

may have gotten out of control, over harvesting fish and animals that live in and around a local estuary, and the construction of middens, which were dump heaps for domestic wastes. Still, in ancient times, nature was largely behind the form and function of human societies; that relationship is best described as geography behind history.

Nevertheless, many ancient peoples made decisions to live in environmentally precarious places like the Roman coastal towns of Herculaneum and Pompeii that were built in the shadows of an active volcano named Mount Vesuvius. In a similar fashion, people choose to live in flood zones and hurricane allies like those which are located on the shores of the Indian Ocean and the Gulf of Mexico. In many famous environmental disasters in history, human actions were behind them; the destruction of Herculean and Pompeii were examples of "history behind geography." If people embraced learning about past geographies, geography behind history, and history behind geography, they might make better decisions about where to live and when and how to shape environmental policies. It is important to recognize that when a hurricane destroys billions of dollars in homes and businesses, there is a ripple effect throughout society as insurance companies raise premiums and governments increase taxes to offset their expenses to pay for the damage. Nearly everyone is indirectly affected by the disaster. Peoples' understanding of the nature and veracity of climate change would likewise increase. In the remaining pages of this chapter, the focus will be on examples of how geography behind history has impacted societies and their use of spaces.

Masada

There are many examples of how weather and climate have impacted history. As with the migration of nations in the medieval days, weather and climate are major environmental factors in forcing societies to alter their cultures and thus their interactions with other societies and with the natural world. From the drying of climate in the eastern Mediterranean to protracted cold epochs in Europe and violent storms and flooding in South Asia and North America, millions of lives have been destroyed, uprooted, or otherwise damaged by weather and climate. While destructive storms are features of weather conditions, they are arguably symptomatic of longer, patterned climate epochs that last for at least 30 years.[9] For instance, tornadoes are produced when aggressive dry air masses collide with bodies of warm, moist air. Those air masses do not randomly appear on the landscape; geographic and climatic conditions have to favor their formation. Currently, the most common place for them to form is in North America where warm, moist air from the Gulf of Mexico penetrates the atmosphere over the Great Plains. In the spring and early summer months, those warm air masses are met by aggressive cold, dry air pushing down from the Rocky Mountains and Canada. A similar situation occurs in the southeastern United States during the spring and fall months. If climate changes, it is possible that the locations and frequencies of tornadic activity would also change. As is shown later, tornadoes appear in the most ancient texts of the Bible, but they are absent from the more recent books of the Old Testament. They are also lacking from the New Testament. Basic landforms (hills, plains, mountains, and valleys) in the eastern Mediterranean have not changed, so there must have been some major shifts in oceanic and atmospheric conditions. Changes in solar output may also have played a role in the shifting frequencies of rain producing frontal activity and tornadoes in the eastern Mediterranean.

FIGURE 1.4 Aerial View of Masada.

© photosounds/Shutterstock.com

---

[9] NASA, *Difference between Weather and Climate*.

© badahos/Shutterstock.com

FIGURE 1.5 The Judean Desert with the Dead Sea in the Distance.

Nonetheless, paleoclimatological (old climates) and historical evidence from the Near East and else-where point to a deep reservoir of climate information which suggest that arid conditions around the Dead Sea, which during the biblical era was known as the Salt Sea or the Eastern Sea, have alternated with wet-ter periods throughout the last 9,000 years.[10] During what must have been a wetter epoch, King Herod of Judea (ca. 74 BC–ca. 4 BC) built a fortress and two palaces on a lofty plateau called Masada. Hydrologist A. T. Wolf argues that during the Roman Warm Period, which included Herod's tenure on the nominal throne, rainfall in the Dead Sea area was 50 percent more than it is today.[11] Nowadays, the high, bone-dry plateau is located in Israel overlooking the Dead Sea from the eastern edge of the Judean Desert. During Herod's time, climbing to the top of the plateau was not easy, so the King thought that the high-level land would make an excellent haven during armed revolts. Masada also offered a tremendous view of the blue waters of the Eastern Sea, which would have contained more water and less saline than it does today. Herod and his would-be defenders constructed storehouses for food and cisterns to collect rainwater; the king's refuge even boasted a bathhouse. As it turned out, Herod never used Masada for its original purpose, but as many as 960 Jews took shelter there after the Romans sacked Jerusalem in 70 AD. While there is some debate among archeologists about how large of a community took refuge on Masada, it must have been big enough to be a concern for the provincial Roman government. The rebels were pursued and besieged by a determined Roman legion for over a year. The Romans, who had earlier occupied Masada as a strategically important place, invested millions of man-hours building a causeway to reach the rebels' sanctuary on the top. When the Romans reached the palaces, they discovered that the rebels had committed mass suicide. However, the number of holdouts on Masada is a debated topic. Kenneth Atkinson, for instance, argues that there is no physical evidence that a large number of people killed themselves on Masada. The remains of only 28 people were found in a cave.[12] However, the Romans could have removed the dead, so they could once again occupy the building and grounds. Clearly, they saw great value in occupying Masada. It makes little sense that the Romans would have allowed the lofty perch to remain a graveyard.

---

[10] Stein, Torfstein, Gavrieli and Yechieli, "Abrupt Aridities and Salt Deposition," 567–575.

[11] Wolf, *Hydropolitics along the Jordan River*, 14.

[12] Sheppard, *The Jewish Revolt*, 83. See Atkinson, *Making History*, 397. See Zias, "Human Skeletal Remains," 732–738.

FIGURE 1.6 Roman Ramp to Masada.

FIGURE 1.7 Van Lake, Turkey.

The regional climate must have been much wetter during the Roman Warm Period (200 BC–400 AD) than it is today. Indeed, studies show that the Dead Sea area was wetter during the history of the early church and the Jewish Diaspora in 70 AD. "Lake Van [the largest lake in Turkey] isotopes indicate a shift away from higher aridity c. 150 b.c. (Figure 7b). The Sofular (a cave in northern Turkey) speleothem (minerals deposited by water in a cave) shows relative stability from about 100 b.c. to about 250 a.d. (Figure 1b). Isotope values in archaeological wood from the Roman siege ramp at Masada may also suggest wetter conditions around the Dead Sea in the first century a.d., which would agree with possible evidence from Ptolemy that Egyptian weather was rainier c. 120 a.d. than today."[13]

---

[13] McCormick et al. "Climate Change during and after the Roman Empire," 169–220, 180.

Nowadays, Masada has a high evaporation rate and only 2 inches (5.8 cm) of rain per year; there is little chance that a handful of rebels could be sustained for a week on today's Masada. In the early to middle 70s AD, the lofty plateau supported anywhere from 28 to 960 people for a year without supplies or water shipped in from family and friends. Masada's example shows how climate can change the way people are able to occupy or use space.

Environmental Push Migration

When weather gets to be too extreme, people migrate to more gentle climates. Despite that historical fact, popular culture in the West has developed a short-sighted, pathological view of colonial European expansion into areas already occupied by indigenous peoples in the Americas, Asia, and Africa. This view assumes that indigenous people were always in those colonized lands and that immigrants invaded their lands because of greed. Because of DNA science, we now know that all men and women on earth descend from a Y-Chromosome Adam and a Mitochondrial Eve who lived in east Africa.[14] That fact alone suggests that each and every one of us descend from immigrants, unless we hail from Ethiopia. In this section, we will take a look at historic migration waves that were at least partially caused by changing climate. Before examples of those migrations are discussed, it is helpful to take a longer look at why people migrate in large numbers. This is important because historians often dismiss underlying environmental and hence climatological factors that exacerbate declining economic and political circumstances that make large numbers of people leave an area.

Why people move from a place (emigrate) is called a push factor, and what attracts them to a place (immigrate) is called a pull factor. Among the important push and pull factors that affect flows of humans are economic, political, environmental, and sometimes geopious.[15] It is important to point out that religious issues might be at the heart of a political push situation, but when the Jews from Europe and the Americas migrated to modern Israel after its formation in 1948, most did so for geopious reasons. Geopiety refers to the devotional or reverential side of religion tied to space. Still, it is sometimes hard to separate out one factor over another in some migration waves. That was certainly the case in the first half of the twentieth century when pogroms in Czarist Russia, later centralizations in the Soviet Union, and the genocidal campaign of the Nazis in occupied Europe pushed millions of Jewish immigrants to the United States. Those waves of immigrants encouraged the newly formed United Nations to recognize a Jewish Israeli state in the eastern Mediterranean.

FIGURE 1.8 Modern Jews at the Wailing Wall.

---

[14] Callaway, "Genetic Adam and Eve Did Not Too Far Apart in Time."

[15] Domosh, Neumann and Price, *Contemporary Human Geography*, 112.

While the dominant push and pull factor throughout the history of humanity is economic, it takes but a little imagination to recognize that weather and climate or environmental conditions played major roles in affecting economic migrations during the hunting and gathering and farming pasts of most societies. If dry conditions impacted plant growth in an area, game animals as well as edible plants would likewise be in short supply. Most people migrated in search of more resources to support their families. It might be handy to appreciate that throughout the history of humanity, survival or economics was at the heart of our ancestors' global conquest. It is also handy to see that in a hunting and gathering or agricultural society, which was how most people lived until the early 1900s, food supplies and hence carrying capacity of an area, was greatly affected by short- and long-term weather patterns. In addition to the indirect effect that weather and climate have on the biological and human carrying capacity of an area, migrations are also caused by having too much water (flooding) or too little water (drought) as well as an in-tolerable range of temperatures.

Formal studies on why people migrate began in 1885 with the publication of Ernst Georg Ravenstein's *The Laws of Migration.*[16] Although he was born in Frankfort in 1834, Ravenstein lived in England for most of his adult life. He was a witness to the Industrial Revolution and observed a great deal of migration in Western Europe and England. While some of the migrants he observed moved internationally, most of them moved intraregionally from rural areas to urban centers. Although a large part of his life was spent in the waning decades of the Little Ice Age, the practice of primogeniture made things worse for farmers' daughters and younger sons. The practice of primogeniture called for the bequeathing of farmlands to the eldest son, so his siblings were often not allowed to continue living on the family farm. The Georgian-era writer Jane Austen (1775–1817) penned at least two books in which primogeniture played major roles in the lives of her characters, especially the daughters.[17]

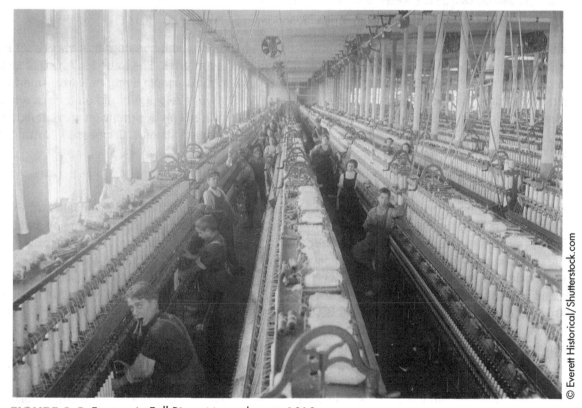

FIGURE 1.9 Factory in Fall River, Massachusetts 1912.

© Everett Historical/Shutterstock.com

---

[16] Ravenstein, *The Laws of Migration.*

[17] See Austen, *Sense and Sensibility* and *Pride and Prejudice.*

FIGURE 1.10 Fans of Jane Austen in Bath, United Kingdom in September 10, 2016.

Restricted growing seasons and primogeniture also played roles in pushing immigrants from Great Britain and Ireland to America during the seventeenth and eighteenth centuries. The Little Ice Age that lasted from ca. 1250 AD to ca. 1850 AD also impacted Algonquin and Iroquois people living in North America. It was during that time that an Algonquin people known as the Shawnee relocated from an area in what is today the New York/Canadian border area to southern Ohio. The Cherokee, a significant tribe among the Iroquois, also left the New York/Canadian border region for the warmer climes of Tennessee, North Carolina, and Georgia during the Little Ice Age.

Climate and Viking Settlements

Looking back further in time to 900 AD to 1300 AD, warmer conditions known as the Medieval Warm Period or Medieval Optimum in the North Atlantic produced a milder climate across northern Germany and Scandinavia.[18] Population pressures were probably already occurring in Scandinavia, but the warmer weather that characterized the Medieval Optimum increased agricultural food production and the amount of time that ships could harvest fish from the sea. In 793 AD, a horde of Vikings sacked Lindisfarne, a monastery on Holy Island in Northumbria (today's northeastern England), and soon found their way to the Hebrides, Isle of Mann, Ireland, France, and even into what is today Russia and Ukraine. Other Vikings ventured across the North Atlantic and were in North America by 1000 AD. Those venturesome settlers had left colonies on Greenland and Iceland. At one time, there were 500 Norse farms operating in Greenland, but when the Little Ice Age settled in over the island in the thirteenth century, most of the farmers vacated their lands. Viking colonies in Iceland fared better than their counterparts on Greenland; today Icelandic, which is spoken by the residents of Iceland, is the closest of the modern languages to Old Norse, the tongue spoken by the original Viking hordes.

---

[18] Mann, "Medieval Climatic Optimum," 514–515.

© noiseshapes/Shutterstock.com

FIGURE 1.11 Viking Hut at L'Anse aux Meadows, Island of Newfoundland, Canada.

Looking Ahead in Climate Change in History

Throughout human history, weather and climate have played major roles in people's decisions to leave places and move to other sites. Sometimes overpopulation in certain areas stretched the capacity of the local environment to support growing villages. Because of our dependency on natural resources, including weather and climate, it is logical that virtually every Mesolithic and Paleolithic society featured an animist religious system. In animism, people believe that plants and inanimate objects found throughout nature have souls. The wind, for instance, has a soul and a spiritual essence. In the 1995 Disney movie *Pocahontas*, animist beliefs are clearly expressed in the song *Colors of the Wind*. Polytheistic and monotheistic religions, too, embrace beliefs that tie together divine worship and attributes with places and the forces of nature. That is the essence of John K. Wright's conception of geotheology.

In the pages that lie ahead, we will take a longer look at the historicity of climate change, the complexity of climate change, and how weather and climate has affected societies. We will then dig deeper into sacred texts penned by Abrahamic faith leaders and see how their descriptions of past weather and climate affected their perceptions of human-environmental relationships. There is much to glean from those texts.

CHAPTER 2

Historicity of Climate Change

INTRODUCTION

When climate researchers look back over time, they hope to establish temporally based estimates for climatic oscillations; as is discussed shortly, no less than two scientists believe that there are 1,300-year breaks between little ice ages. However, among the many complicating factors associated with establishing global climate change patterns is the regional setting which may deviate from the larger pattern. For instance, much has been made in the climate change debate about the decline in arctic ice between 1979 and 2012, but what is often dismissed in the discussion is that there are measurable increases in Antarctic ice over the same time period.[1] Logic suggests that if warming is causing a decrease in arctic ice, cooling must be increasing ice cover in and around Antarctica. As is shown later in the chapter, there is some debate among scientists about the regional nature of what is otherwise thought to be global climate change.

As I noted in *The Forces of Nature*, "One important fact is revealed by looking back through the pages of history and through the geologic record: climate change happens, and it will continue into the future. Ironically, higher levels of atmospheric water and carbon dioxide may actually keep the earth's climate from slipping back into a Pleistocene-like epoch. Not only would a recurrence of an ice age spell certain doom for European cities like Moscow, Berlin, Geneva, Oslo, Manchester, and Birmingham, among others, North American cities like Montreal, Toronto, Minneapolis, Lansing, Buffalo, Lincoln, and Albany would suffer the same destructive fate."[2] In such a scenario, North America's biological carrying capacity over a huge section of the continent's rich breadbasket would be greatly reduced or lost; in the southern portion of the American breadbasket, the growing season would be shortened. Given current technology and the world's dependence on American and Canadian agricultural exports, both nation's collective ability to feed hundreds of millions of people would be reduced. With respect to Canada, the country is the fifth leading exporter of agricultural products, and some 12 percent of the Canadian workforce is employed in agricultural and related industries.[3] The return of even a Little Ice Age would be a devastating hit on the Canadian economy and shift the burden to feed millions of people to other societies which may not have the infrastructure to fill the void in food production. The return of another ice age is possible. As we will see in Chapter 3, many informed climate watchers in the 1970s were convinced that a return of an ice age was imminent. This begs the question: can the recurrence of an ice age be predicted? As is shown in the next section, climatic patterns since the Roman era are not very precise.

---

[1] NASA, "Arctic and Antarctic Sea Ice."

[2] Vann, *The Forces of Nature*, 231–232.

[3] AGRI-LMI Labour Market Information, "Canada's Agriculture Sector Labour Market."

FIGURE 2.1 Farming in Corn Hill, Canada.

FIGURE 2.2 This South Dakota Farmland would be Lost to Production if an Ice Age Returned.

The Historicity of Climate Change

As mentioned above, higher levels of greenhouse gases like humidity, nitrous oxide, ozone, carbon dioxide, and methane could save humanity from a nasty, cold, if not fatal, future. To fully appreciate that thought, we need to question the utility of the conventional view of dynamic equilibrium (nature equitably recycles elements like carbon) when it is applied to recycling carbon and water, the two major greenhouse culprits in global warming. Clearly, nature recycles water and carbon, but the notion that nature balances the amount of carbon and water that are added to and removed from the biosphere is not supported by physical evidence. Natural systems, if left to themselves, will remove and store vast amounts of water and carbon from circulation, which, in the long-term, would have a cooling effect on weather and climate and could help explain why there have been eight different "ice ages" in the past 720,000 years.[4]

In the case of water, sandstone aquifers such as the Oglala, which lies beneath the western plains of the United States, once held trillions of gallons of water; nature removed that water from the hydrologic cycle. However, mining water to irrigate range and croplands as well as to supply municipalities with potable water is returning significant amounts of the resource to the hydrologic cycle. Also, alpine and polar glaciers retain water that has not been in gaseous form for millennia (232).

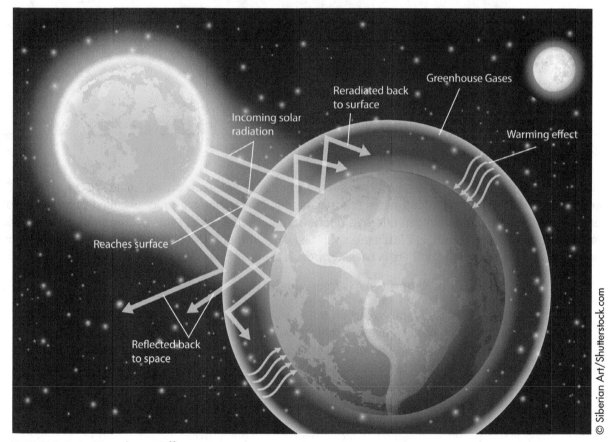

FIGURE 2.3 Greenhouse Effect.

© Siberian Art/Shutterstock.com

---

[4] Penna, "Climate Change," 579.

© NosorogUA/Shutterstock.com

FIGURE 2.4 Aquifer Containing Groundwater.

Carbon, alternatively, begins its journey in supporting life as gaseous carbon dioxide (CO_2). When CO_2 enters a plant through its breathing pores or stomata, sunlight initiates a process called photolysis that breaks water molecules into oxygen and hydrogen. Carbon is then fixed with hydrogen to form sugars. Oxygen is an important byproduct of this process. It escapes through the plant's breathing pores. Carbon produced and contained in plant fibers is passed through the food chain (233).

As I wrote in *The Forces of Nature*, (This is from page 233)

Like water, natural processes remove carbon from its cycle. In addition to trapping carbon in coral reefs, billions of tiny globigerinas live out their short lives in the world's oceans. Their remains fall from the surface of the seas like an organic rain, and their calcium-carbon-based skeletons accumulate on the ocean floor. Ralph Buchsbaum estimated that about thirty percent of the ocean floor (some 40 million square miles) is covered in grey globigerina ooze. In deeper parts of the ocean, the ooze eventually forms chalk and in shallower coastal areas, the ooze lithifies into limestone. Clearly the formation of coral reefs and the buildup of chalk and limestone layers provide evidence that carbon is not always recycled and used in the circle of life. Further evidence of this process is shown in the formation of coal and crude oil deposits.[5]

---

[5] Vann, *Forces of Nature*, 233.

FIGURE 2.5 Limestone Rock.

© Salienko Evgenii/Shutterstock.com

© Ekaterina Pokrovsky/Shutterstock.com

FIGURE 2.6 Chalk Cliffs of Étretat, Normandy, France.

The Paleozoic Era contained a large section known as the Carboniferous Age. It lasted from about 290 to 354 million years ago. During that time period, the earth's climate was warmer and wetter. Lush forests, including giant tree ferns, grew in much of the eastern portion of North America, especially in Ohio, Pennsylvania, Virginia, West Virginia, Kentucky, Indiana, and Illinois as well as across central and southern England and central Europe. That was also a time of tectonic mountain building caused by continental collisions. In North America, the period is divided into the Lower and Upper Carboniferous Eras. Whereas the Upper Carboniferous is identified by mostly coal deposits, the Lower Carboniferous is associated with limestone, a sedimentary rock composed of calcium carbonates deposited on ancient sea floors by the remains of crustaceans and crinoids (lime encrusted algae). The Upper Carboniferous Age had a tropical climate that supported vegetation that is similar to the plants found in the Orinoco Delta of eastern Venezuela[6] (233).

---

[6] Scheihing and Pfefferkorn, "Taphonomy of Land Plants," 205–240.

FIGURE 2.7 Tropical Forest in the Orinoco Delta, Venezuela.

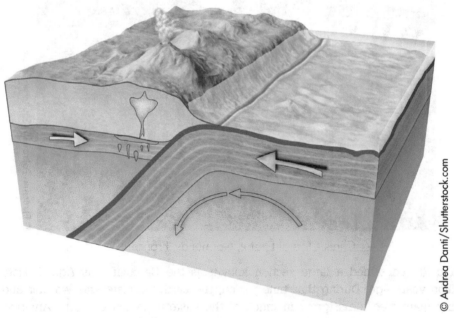

FIGURE 2.8 Continental Plate Tectonics.

The effects of continental or plate collisions must have been dramatic because sediments quickly covered large quantities of carbon-based plant matter. In doing so, huge quantities of carbon were removed from the circle of life. In more scientific terms, a significant portion of the elemental cycle was removed from the biosphere. Those important events in the earth's natural history hints at a process that continued through the ages and likely led to the Pleistocene, the last great ice age. By removing huge quantities of carbon dioxide and humidity from the biosphere, nature cooled itself and killed off countless species of plants and animals that could not adapt (233–234).

The Holocene, the relatively warm interglacial period in which we find ourselves, is only about 10,000 years old. The Holocene is also known as the "Age of Man" because all of recorded history has occurred during this time period and because human population climbed from less than five million to more than seven billion. Some scholars, who are following the lead of Nobel-Prize winning atmospheric chemist Paul Crutzen, contend that we are currently in the Anthropocene (a human-caused climate epoch). However, that contention is challenged by Whitney Autin, a stratigrapher working at CUNY Brockport, who claims that the Anthropocene is more about appealing to pop culture than it is supported by science.[7] Nevertheless, if the Holocene's proximity to advances of continental glaciers is correct, the earth's climate will return to a much colder state. However, there is no way of knowing when or if another ice age will occur. Nevertheless, a return of an Ice Age would be devastating to humanity. With a drastic reduction in the earth's culturally-modified carrying capacity, social order would collapse and anarchy would be activated with unimaginable consequences. Aspects of this unsavory topic are, for the time being, perhaps mitigated by human actions, so let us return to a temporal delineation of climate change and how releasing carbon dioxide and humidity back into the atmosphere may not be such a bad thing. This is important to consider because most of the contemporary thinking on climate change has focused on warming as a bad thing. Such thinking is not likely to change anytime soon because the year 2009 ended the warmest decade on record[8] (234).

Since 1880, the year in which reliable thermometers were put into use at various places around the planet, global temperatures have risen by 8°C (1.5°F).[9] However, between 1940 and 1980 temperatures declined almost to their 1880 level, and as was shown above, winter storms in December of 2010 as well as January 2011 dumped record breaking snow falls on Connecticut and Manhattan and created havoc and economic disruption throughout the Northeast.[10] The blizzard of Christmas weekend in Philadelphia, Pennsylvania, was so severe, that an NFL football game scheduled for December 26, 2010 was cancelled. Mother Nature was trying to wreak havoc on the NFL.[11] The massive snowfall and associated cold weather was not limited to the Northeast (234–235). Earlier in the month on the night of December 12-13, some 18-20 inches fell on Minneapolis, Minnesota, a city made partially famous for its innovative Mall of America and the "indoor effect" it has created in the downtown portion of the city. Among the city's innovations was a dome, handily named the "Metrodome," that was built for its Vikings football team. The fiberglass, Teflon coated roof collapsed under the weight of 18 inches of snow.[12] Fortunately, it was not during game time. Admittedly, the snowy events that occurred in successive winters in 2010 and 2011 do not indicate that an ice age is returning, but they do suggest that weather patterns are not linear (235), especially when one considers that record snow fall blanketed New York City again in January 2016: "Snow fall totals in Central Park were upped from 26.8 inches to 27.5 inches, making the Jan. 22-23 storm the biggest blizzard to hit the city since record keeping began in 1869, according to the National Oceanic and Atmospheric Administration."[13] Change in weather and climate does happen. A longer look at the earth's meteorological past supports that contention.

---

[7] Stromberg, "The Age of Humans."

[8] Goddard Institute for Space Studies, "Second Warmest Year."

[9] Ibid.

[10] Trotta, "New York Hard hit"; NOAA, "National Snow and Ice."

[11] ESPN. "Fierce Storm Sacks."

[12] Carty, "Metrodome Roof Collapse."

[13] NBC New York, "NYC's Biggest Snowstorm."

FIGURE 2.9 Snowy Central Park in March.

FIGURE 2.10 Hubert H. Humphrey Metrodome in Minneapolis, Minnesota.

As is shown later in Chapter 6, imagery of weather and climate in the Bible suggests that conditions in the Near East were cooler during the time in which a number of history books in the Bible were written (ca. 950 BCE). Indeed, during the time of Abraham and Lot (2100 BC–2000 BC), the plains of the Jordan River and today's Dead Sea were much wetter than they are today and were able to support growing villages.[14] Non-biblical evidence for cooler climate conditions during the writing of the Bible's history books is suggested in the research conducted by Lloyd Keigwin who studied salinity and the flux of terrigenous material (weathered rock from continental sources) in the Saragossa Sea, which despite being

[14] Migowski et al., "Holocene Variability and Cultural Evolution," 421–431.

situated west of the Mediterranean Basin in the middle of the North Atlantic, nonetheless shares a similar latitudinal location as the northern Near East. Keigwin found that temperatures prior to medieval times were 1°C (1.8°F) cooler than they were in 1996.[15] In a similar manner, non-biblical support for icy conditions in the Near East during that time period is provided by paleoclimatologists Charles A. Perry and Kenneth J. Hsu who creatively tied together geophysical, archeological, and historical evidence to support a solar-output model for climate change. Their study identified 1,300-year cycles of climatic cooling and heating during the Holocene. They contend that the 1,300-year cycle for little ice ages was only skipped over once, and that was around 7,500 BCE.[16] As is shown in a later chapter, the people who wrote the history books in the Old Testament were clearly witnessing a cold climate. Perry and Hsu describe in good detail a little ice age that lasted from 1250 BCE to 750 BCE, a time frame that most likely coincided with the writing of those books. Assigning a date for the Book of Job, however, is difficult, but theologians and bible editors Earl D. Radmacher, Ronald B. Allen, and H. Wayne House, argue that the mention of iron tools (19:24, 20:24, and 40:18) suggests a date after 1200 BCE. More specifically in Job 7:17, they point out that the author seems to allude to a passage in Psalms (8:4), so it is quite possible that Job was written during the time of Solomon (ca. 1017–931 BCE)[17] (235–236). However, as is shown later, weather patterns in the book of Job are quite different from the atmospheric conditions in the history books and the most recent Old Testament texts. The distinctly cold and volatile weather patterns depicted in Job suggest an older authorship.

FIGURE 2.11 Jordan River, Israel.

---

[15] Keigwin, "The Little Ice Age," 1503–1508.

[16] Perry and Hsu argue that solar activity is the major cause of climate change. See their "Geophysical, Archaeological, and Historical Evidence," 12433–12438.

[17] Radmacher, Allen and Wayne House are theology professors at Western Conservative Baptist Seminary, Dallas Theological Seminary, and Michigan Theological Seminary, respectively. They are also Bible editors for Thomas Nelson, Inc. Their analysis of the Book of Job is found on page 769 of the NKLV Study Bible Second Edition (Nashville: Thomas Nelson, 2007).

Perry and Hsu go on to argue that a little ice age that befell the planet between 60 AD and 600 AD was an environmental force behind what they call the "migration of nations."[18] However, studies in the eastern Mediterranean point to a much warmer climate that is called the Roman Warm Period, which suggests that, as is the case today with growing glacial ice in Antarctica and shrinking ice in the Polar regions, regional climate patterns often defy global patterns.[19] In particular, they point out that it was during this time that ancient Germanic peoples such as the Vandals, Goths, and Asiatic folk like the Huns besieged the Roman Empire. In Asia, the Chinese Empire was likewise overwhelmed by Asiatic peoples. On the other hand, this little ice age helped Central America's Mayan civilization flourish on the Yucatan Peninsula.[20] Whereas Germanic and Asiatic tribes migrated to warmer environs where they ran into conflict with existing civilizations, the cooler and drier weather in the Yucatan drove malaria carrying mosquitoes further south. The tropical climate gave way to a more temperate weather pattern that was better suited to cultivation and city building. However, when the Medieval Warm period (also known as the Medieval Optimum) arrived, so did the mosquitoes that carried the scourge of malaria.

The Medieval Warm period which ran from ca. 800 to 1250 is well-documented in Scandinavian migration history, but in the scientific literature its existence has been disputed by some scholars while others dismissed it as a minor regional phenomenon.[21] However, as scientists from the southern hemisphere have joined in the quest to challenge the historicity of the Medieval Warm period, it is clear that it occurred in both the northern and southern hemispheres.[22] Those studies challenge the veracity of the sweeping and dismissive conclusions made by Werner et al.

The Little Ice Age that ran from 1280 to 1860 was perhaps the coldest period on earth since the end of the Younger Dryas about 11,500 years ago.[23] The Little Ice Age was also a time of the Bubonic Plague and the resulting Black Death that swept across Europe and arrived in the British Isles in 1348 (236). The River Thames routinely froze and was sufficiently safe enough for people to skate on it. In Europe, alpine glaciers extended well into valleys inhabited by people. In the Americas, the Cherokee, who spoke an Iroquoian language, migrated into southern Appalachia from the St. Lawrence River area of upstate New York and southeastern Canada. The Little Ice Age was also arguably responsible for the migration of the Shawnee into southern Ohio and Kentucky. The Shawnee spoke an Algonquian language and were cultural relatives of the Algonquin tribe of Quebec and Ontario. While there is no written record of the reasons for their relocation, the Cherokee obviously left behind their Iroquois relatives in the Northeast to resettle in what are today North Carolina, Tennessee, and Georgia. Given the fact that the Little Ice Age had reduced the regional carrying capacity in eastern Canada and upstate New York, it seems logical to conclude that their migration was pursued for the sake of survival. The settlement of the Americas by Europeans obviously occurred during the Little Ice Age. The colder climate precipitated even Irish Sea migrations between Britain and Ireland[24] (236–237).

Temperatures during the Little Ice Age would have varied in accordance with latitude, elevation, and proximity to large water bodies. Locations on the windward side of a continent or large island like Greenland, for instance, would have seen less of a change than places located on the leeward side of the island or continent. Tropical places with low elevations like the Yucatan Peninsula in modern-day

---

[18] Perry and Hsu, "Solar-Output Model," 12435.

[19] Werner et al., "Globally Coherent Warm," 550–554.

[20] Perry and Hsu, "Geophysical, Archeological, and Historical Evidence," 12435.

[21] Crowley and Lowery, "Medieval Warm Period," 51–54; Hughes and Diaz, "Medieval Warm Period," 109–142.

[22] Villalba, "Tree Ring and Glacial Evidence," 183–197.

[23] Perry and Hsu, "Geophysical, Archeological, and Historical Evidence," 12435.

[24] Vann, *Search of Ulster Scots Land.*

© Lillian16/Shutterstock.com

FIGURE 2.12 Skara Brae, Scotland.

Mexico would have enjoyed a temperate climate that may well have resembled Mount Sandel (7,000 BC) in Ireland or Skara Brae (3,000 BC) in the Orkneys during ancient and long forgotten warm periods.[25] On balance, however, the average temperature in the North Atlantic, according to Lloyd D. Keigwin, who studied geophysical evidence in the Saragossa Sea, was 1°C (1.8°F) lower than it was in the 1990s.[26] Again, this needs to be seen in light of the fact that the range of temperatures is always lower in marine climate zones than they are in continental locations like the Great Plains or Siberia (237).

Studies conducted in South America's Patagonia show that there was a cold, moist climate in the region from 1270 to 1660.[27] Further to the west in the South Pacific, Erica J. Hendy and a host of her colleagues conducted a study that delineated a 420-year history of strontium/calcium, uranium/calcium, and oxygen isotope ratios measured in eight coral core samples taken from the Great Barrier Reef located off of the northeast coast of Australia in the Coral Sea. The results of their extensive study showed that salinity was higher in the mid-1500s through the middle of the nineteenth century than it was in the twentieth century. Hendy and her team concluded that the sea "freshened" because warmer temperatures toward the poles after 1870 reduced the velocity of winds over the tropics and hence lowered the amount of moisture evaporated from the waters off of the coast of Australia. As a result of lower evaporation rates, they reasoned, more water was available to dilute sea salt in the tropical latitudes of the South Pacific[28] (237–238).

---

[25] The dates of occupation for Skara Brae and Mount Sandel fit into the warm periods identified by Perry and Hsu, "Geophysical, Archeological, and Historical Evidence," 12435–12436.

[26] Keigwin, "Little Ice Age and the Medieval Warm Period," 1506.

[27] Villalba, "Tree Ring and Glacial Evidence," 183.

[28] Hendy et al., "Abrupt Decrease in Tropical Pacific Sea Surface," 1511–1514.

FIGURE 2.13 Great Barrier Reef of the Northeast Coast of Australia.

FIGURE 2.14 Krakatoa, Indonesia.

Between 1880 and 1940, global temperatures rose slightly above the world-wide average temperature of 59°F. A gigantic eruption on the East Indian island of Krakatoa in 1883 spewed particulate matter into the atmosphere from May 20th until August. Sunlight striking airborne matter from Krakatoa was reflected back into space, and, as a result, global temperatures were markedly lower throughout the next year. Its effects on weather were strong in the northern hemisphere. Bostonians were able to throw snowballs in June. Nevertheless, the phase of warming reached its maximum effect during the late 1930s during which time over grazing and deep-plow cultivation in the western Great Plains contributed to the Dust Bowl (238).

FIGURE 2.15 Mount Pinatubo Crater, Philippines.

From around 1940 through the late 1970s, global temperatures declined to just below the global average.[29] Global temperatures dropped to a near record low in the wake of the eruption of Bali's Mount Agung in 1963.[30] Almost thirty years later on July 16, 1990, the island of Luzon in the Philippines was rocked by an earthquake that registered 7.8 on the Richter scale. The epicenter was sixty miles northeast of Mount Pinatubo. The earthquake was a precursor to a build-up of volcanic gases that caused a violent eruption of Pinatubo on June 15th. The gaseous explosion expelled one cubic mile of sunlight deflecting material into the atmosphere. Throughout the remainder of 1991 and on through 1993, global temperatures dropped by 5°C (about 1°F).[31] Thanks in large measure to the predictions made by the Philippine Institute of Volcanology and Seismology and the US Geological Survey, at least 5,000 lives were saved and $250 million in property damage was averted.[32] While the US Geological Survey and writers like Oliver S. Owen give ample attention to the cooling effects of particulate matter spewed from volcanic eruptions, they seldom note the possible climatic connection to the huge volume of greenhouse gases that are belched out into the atmosphere. The most abundant gases usually released in an eruption are H_2O followed by CO_2 and then by Sulphur dioxide (SO_2).[33] Particulate matter settles out of the atmosphere long before nature can down-cycle sudden bursts of CO_2 from eruptions like Pinatubo. Given the fact that the prevailing weather pattern from 1980 to 2009 was one of increasing temperatures that reached a peak in the first decade of the twenty-first century, the role played by greenhouse gases emitted by volcanic eruptions in global warming scenarios deserves further study (238–239)..

---

[29] Owen, *Natural Resource Conservation*, 579.

[30] Ibid.

[31] United States Geological Survey, "Cataclysmic 1991 Eruption of Mount Pinatubo."

[32] Ibid.

[33] Symonds et al., "Volatiles in Magmas," 1–66.

FIGURE 2.16 Snowy Scotland.

December 2009 ended the hottest decade on record, but climate could cool down again. Indeed, as is shown in the chapter on the complexity of climate change, weather conditions in the winter months of 2010 and 2011 took a nose dive. As mentioned above, cities like New York and Boston in the Northeast were not inundated by rising sea levels caused by melting ice caps and glaciers, but they were hard hit by successive snow storms and blizzards. The winter rampage was not limited to the Northeast and the contiguous states, however. In the United Kingdom, 2010 witnessed the coldest December on record. However, record-keeping on a national scale in the UK only goes back to 1910.[34] Nonetheless, December 2010 was not just a little colder than usual. The long-term average for the month is 4.2°C (39.56°F), but it dropped to 1°C (32.18°F) in 2010. Elsewhere in the UK, December temperatures were at an all-time low in Northern Ireland, and they were at their lowest in Scotland since 1947[35] (239).

While these events do not necessarily suggest that a mini or great ice age is looming, they should remind us that predicting climate change is fraught with a host of problems. It should also remind us that the return of a major ice age could happen, but, as research by Perry and Hsu suggest, it is more probable that a little ice age will reappear sometime during the next millennium. However, given that we live in the Holocene, which is an interglacial period, a great ice age is due to return. It is a paradox to think that human activities, so lamented by those who fear climatic geokolasis (earth or nature punishes), may actually keep mid-latitude expanses of the earth habitable for centuries to come (239–241).

[34] Independent, "2010 Showed Coldest December on Record."

[35] Met Office, "Snow and Low Temperatures."

© Girl Grace/Shutterstock.com

FIGURE 2.17 County Antrim, Northern Ireland.

SUMMARY

The historical nature of climate change and the tendency for climate over the past million years or so to move back and forth between "ice ages" and "interglacial warm periods" should get more attention than it receives in academic and popular literatures. Global warming has dominated the discourse and the geokolasis fears associated with it has caused some proponents of human-caused climate change to call for radical changes in economic activities and even control of individual and collective behaviors. Adopting the term Anthropocene to describe our current climate epoch is not useful in looking at climate patterns over the millennia because we do not have temperature data from before the 1870s. It follows that many members of the public and the academy are willing to sacrifice personal liberties to protect the planet from our exploitive activities. As suggested by cultural geographer Mike Hulme, proposals for governance on a global scale are brash, utopian, and not very practical.[36] Indeed, they are little more than examples of the historic tendency of people to offer repentance for sinful behaviors that invited the wrath of nature to befall wayward humanity. Whereas our ancient and recent ancestors saw these scenarios as geotheokolasis (God uses nature to punish) events, more secular people today have expunged God, but they still maintain geokolasis (earth or nature punishes) beliefs.

---

[36] Hulme, "The Conquering of Climate," 5–16, see specifically 5.

As we look ahead, our focus should include contingencies for a return of an ice age. Climate change presents many challenges, but, as is shown in Chapter 8, all contingencies for coping with these and other formidable forces of nature need to be made with respect to the free participation of individuals, groups, and societies. Our survival as a species has always depended on adaptations made by micro and macro social organizations. Just as our Federal system allows for 50 distinct experiments in democracy to try-out novel ideas, the world's 194 countries and their communities and industries offer the best chance for identifying ways to adapt to challenges of climate change. However, not all countries share the same world view and hold an equal understanding of the state of our ecumene (241–242).

CHAPTER

3

The Complexity of Climate Change

INTRODUCTION

To unravel the historical nature of climate change, especially in regard to weather and climate narratives found in Scripture, it is helpful to appreciate, if not understand, the complexity of climate change. In and of itself weather, which is short-term atmospheric conditions, is greatly affected by natural heating and cooling processes and their association with air flow and ocean currents, which disperse heat poleward. Climate is a long-term weather pattern, so climate change is a function of persistent changes in atmospheric conditions over extended periods of time. While some observers of long-term weather patterns contend that humans are the main cause of variations in atmospheric conditions, serious scientists understand that nature itself plays a major role in any changes made in climate.

Nature's involvement in climate change scenarios occurs in albedo ratings (reflectivity of natural objects and surfaces), the elemental cycle, hydrologic cycle, earth's tilt, solar output, and volcanic eruptions. Human activities, such as burning carbon containing fuel, no doubt also play a role in long-term weather patterns. However, keep in mind that nature stores carbon and removes it from the recycling process that scientists call dynamic equilibrium. The challenge to delineate how much of a role that humanity plays in global climate conditions is a serious one. By the same token, it is equally hard to determine how much nature itself affects long-term weather, although Swiss scientists claim that it is now possible to quantify solar effects in weather change scenarios.[1] While delving into how those natural processes affect climate, let us take a look at the factors that impact climate in the context of changes in weather patterns since 1940. By doing so, we can catch a glimpse of the complexity of climate change.

Factors in Climate Change

For anyone living in the United States, Canada, or Europe, it is hard to escape hearing or seeing news and political arguments about climate change. Most of the argument is over human culpability and how we should respond to varying weather conditions. Interestingly, the use of the label "climate change" was not always in vogue. Back in the 1980s, for instance, people heard claims that the earth was experiencing "global warming," and the number one culprit in the warming of the environment was humanity. Like our earliest ancestors who believed that environmental catastrophes were the

---

[1] Swiss National Science Foundation, "Sun's Impact on Climate Change."

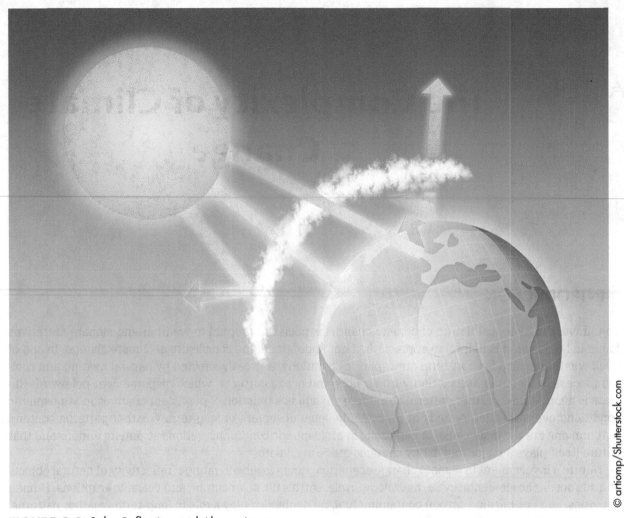

FIGURE 3.1 Solar Reflection and Absorption.

results of judgments of angry gods whose human adherents failed to live up to some standard, modern-day climate alarmists fear that an aggrieved planet is venting its wrath on greedy, wasteful humanity. Aside from the superstitious aspects of religion, little has changed in our reaction to adverse weather. Somehow, we humans have always blamed ourselves for whatever befalls us. The current fear is based on a theoretical model tied to the use of fossil fuels. Specifically, our consumption of carbon-based fuels releases colorless carbon dioxide (CO_2) into the atmosphere where the gas traps longwave radiation. Longwave radiation is also called sensible heat because it can be felt or sensed, and it can be measured on a thermometer. By increasing the level of CO_2 in the atmosphere, humans, as the theory explains, are contributing to what has been called the greenhouse effect. If you have ever gotten into a car with its windows rolled up on a sunny, summer day, you have direct experience with the greenhouse effect. Sensible heat is trapped inside the automobile, making it stifling hot for passengers. In a similar fashion, greenhouse gases trap sensible heat in the troposphere, the lower layer of the earth's atmosphere where most forms of life exist.

Greenhouse effect

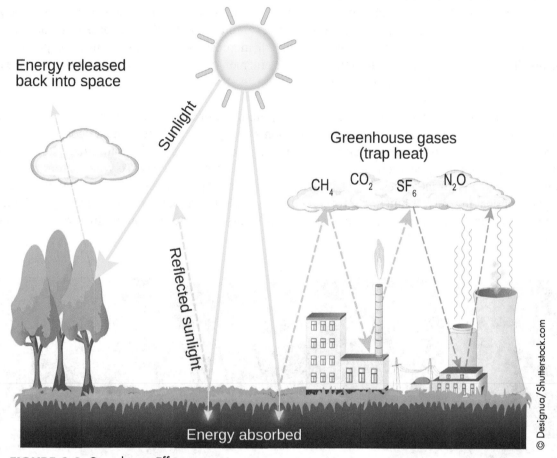

FIGURE 3.2 Greenhouse Effect.

CO$_2$, however, is not the only greenhouse gas; nitrous oxide, methane, and humidity also trap heat in the troposphere. Interestingly, few observers blame variations in humidity on climate change, despite the fact that people play a significant role in reintroducing nature's storehouses of water back into the hydrologic cycle. In many parts of world, including in the western high plains of the United States, people depend on harvesting ancient groundwater that nature stored away in mostly sandstone aquifers. By pumping groundwater to the surface and into homes, businesses, ranches, and farms, humans are returning water to the hydrological cycle. You might recall that the hydrologic cycle consists of three principle phases: condensation, precipitation, and evaporation. Sublimation occurs when ice or snow skip a liquid state on their way to becoming humidity. In gaseous form, H$_2$O (humidity) is a greenhouse gas, and it routinely affects our comfort levels. To consider how humidity affects us, imagine you are in a lower-middle latitudes desert, like the Painted Desert in Arizona, where humidity is often less than 10 percent. It can easily reach 110°F on a summer day, but at night, temperatures may drop to 60°F or less as sensible heat is radiated or lost to outer space. When our sweat or any water evaporates, the process transforms sensible heat into latent energy, so we "sense" this transition in water's state as an energy loss or a cooling effect. If the temperature and humidity are high, on the other hand, the evaporation rate is low, so we would not be very cool, even at night. To contemplate that sensation, imagine that we are in the swampy backwoods of Louisiana in mid-July. It is so humid that our sweat barely dries, leaving our skin clammy and sticky. The swamp's daytime high temperature of 95°F may only drop to around 80°F in the dark of a short, summer night. As the air temperature drops, relative humidity will increase, which in turn, causes evaporation rates to decline. It can be easily seen (and felt) that atmospheric moisture or humidity impacts the way we sense and measure heat. It traps longwave radiation while it lowers evaporation rates, which in turn reduces the cooling effect we experience.

Just as nature stores away water in glaciers and ground water, natural processes likewise store away carbon. Besides coal, petroleum, and peat, the environment captures carbon in chalk beds, limestone rock, and coral reefs. It is important to remember that we are not importing CO_2 from other planets; we are simply putting back into circulation an element that nature has stored away. To what extent that anthropogenic (human-caused) increases in CO_2 play a role in changing weather patterns remains an elusive, yet highly political mystery. If humans returned all of the carbon that nature stored away in coal and peat deposits, the climate would most like resemble the Carboniferous Age that lasted from ca. 359.2 and 299 million years ago.

FIGURE 3.3 A Humid Louisiana Swamp.

FIGURE 3.4 Painted Desert, Arizona.

Still, most of the public's concern about CO_2 build-up in the atmosphere is rightly on increasing air temperatures. According to Columbia University professors James Hansen and Makiko Sato, as well as NASA's Goddard Institute's Redo Ruedy, there is strong evidence that the frequency of extremely hot temperatures around the world has increased since 1981 and will likely continue to rise. As Hansen et al. argue, "The decade-by-decade shift to the right of the temperature anomaly frequency distribution will continue, because Earth is out of energy balance, more solar energy absorbed than heat radiation emitted to space, and it is this imbalance that drives the planet to higher temperatures."[2] It is important to note that for this scenario to unfold, more solar radiation or insolation will need to be absorbed in the biosphere and re-radiated back into the atmosphere as long-wave radiation, or high levels of greenhouse gases will need to be present to retain sensible heat. The model developed by Hansen et al. does not consider changes in humidity. In spite of these factual and perceptional contexts, the media and activists now call changes in weather conditions "climate change," not "global warming." The use of the new label invites alarmists to refer to cold snaps as evidences of human-caused climate change. This change in labeling adverse weather should be a concern to all of us because it invites reactionary government officials to enact legislation or policies that may not be based on good science.

As if the sun's production of energy is perfectly constant, many concerned observers see any long-term weather change as the result of human activities when nature, as was shown above, plays a significant role in climate change. Even minor changes in the amount of energy produced by the sun can create substantial, even life threatening, changes in weather and climate here on earth. In 2017, the Swiss National Science Foundation published a report in which it is claimed that scientists have, for the first time, quantified the sun's impact on climate. The report acknowledges the existence of changes in solar activity and solar cycles.[3] There is, however, danger in assigning specific time frames to the frequency and duration of solar

---

[2] Hansen et al., *Public Perception of Climate Change*, 2.

[3] Swiss National Science Foundation, "Sun's Impact on Climate Change."

phases or cycles. I am not convinced that the sun's activities can be precisely or even imprecisely timed. Still, our primary terrestrial energy source is the sun, and it is amazing that the amount of energy the sun emits is relatively uniform; however, as mentioned above, even modest changes in the amount of energy the sun produces can have serious impacts on our weather. Climate too can be affected if the variation in sun-energy generation is protracted or restricted for long periods of time.

For instance, recall the solar storms of March 5 to 7, 2012.[4] As the calendar would suggest, North America was still in the throes of late winter. At the time, I lived 25 miles north of the Tennessee and Kentucky border. In early March, a radio announcer, who was also an airplane pilot, told his audience to expect radio wave interference from the looming solar storms. Not only were there detectable radio wave interferences, which impacted air traffic communication as well as his broadcasts, there were also changes in our late winter and early spring weather. It was unseasonably warm with mid-summer like heat across the South during the second half of March and April. In considering the effects of a major solar storm, some have theorized that if a solar event such as the one that occurred on August 28, 1859 happened again, the impacts on satellites and electronic communication devices would be catastrophic.[5] It is perhaps no coincidence that 1859 came at the end of the Little Ice Age (ca. 1280–1850). Increased solar activity may well have played a role in the demise of that protracted cold epoch. Such an event almost occurred on July 23, 2012 when an extreme solar storm just missed the earth. The destructive solar convulsion passed through the earth's orbital path just nine days after we passed the area. Source: NASA Science (July 23, 2014). "Near Miss: Solar Superstorm of July 2012. https://science.nasa.gov/science-news/science-at-nasa/2014/23jul_superstorm

FIGURE 3.5. Extreme Solar Storm.

[4] NASA, "Sunspot and Flare."

[5] Mason et al., *Physical Geography*, 60.

Nonetheless, it is perhaps psychologically beneficial in the short-run if people think we are responsible for what befalls us. If we are culpable, we can change our behaviors, but we cannot change the amount of energy that the sun produces or the greenhouse gases the earth emits. How would the public react if it knew how much damage the sun could inflict on our fragile environment? At least for the time being, humanity, especially members of the political left, seem content with assuming total responsibility for any kind of climate change that occurs. Ironically, particulate pollution in the upper atmosphere could actually protect our habitat from too much solar energy. Cleaning the air of particulates could exacerbate the present weather scenario by allowing more sunlight absorption in lakes, seas, lands, and cultural structures like asphalt parking lots and highways.

© digis/Shutterstock.com

FIGURE 3.6 Sunny, Tropical Sky Heats Oceans and the World.

FIGURE 3.7 Low Albedo of Dark Surfaces Like This Highway Enhance Heat Generation.

In a popular culture context, the climatological paradigm shift from "global warming" to "climate change" can be seen in Hollywood movies. Some postapocalyptic films produced since 1990 have been based on such random weather events as advancing ice ages to a flooded, warm-earth scenario. In either case, these postapocalyptic worlds serve as fictionalized evidences of humanity's culpability in changing weather patterns. Movies like *The Day After Tomorrow* and *Water World* come to mind when thinking about the emerging cataclysmic continuum that ranges from excessive heat to killer cold.[6] On the other hand, the film *Knowing* showed the futility of human efforts in the face of a massive solar storm.[7]

Another reason for finding it hard to calculate how much humans are impacting global climate is the variation in cloud cover over tropical oceans. Ocean currents and air currents, including hurricanes, are two of the ways through which nature redistributes excessive long-wave energy from zones of surplus (40° N–40° S) into regions of energy deficits (90° N–40° N and 90°S–40°S).[8] Casual observers of weather events fail to consider the role that hurricanes play in contributing to global temperatures; they typically think that hurricane intensity and frequency are produced by human-caused global warming when the sun and cloud cover are more likely the cause. For instance, after Hurricane Katrina drenched Florida and devastated the Gulf Coast of the United States in August 2005, many news agencies were quick to blame the intensity of the storm on human activates and predicted that more and even deadlier hurricanes were likely to occur. The year 2005 was indeed active with hurricanes. Florida was visited by Katrina in August, Rita made landfall on September 20, and Wilma exacted a toll on the Sunshine state in late October. However, because the dispersal of energy from overheated waters and the lengthy periods of time in which clouds either absorbed or deflected solar radiation back into space, those dire predictions thankfully did not materialize. It was another 11 years before Hermine, a category one hurricane, made landfall in Florida on September 2, 2016.

---

[6] Emmerich, *Day after Tomorrow*; Reynolds, *Water World*.

[7] Proyas, *Knowing*.

[8] Mason et al. *Physical Geography*, 64.

summer solstice (June 21)

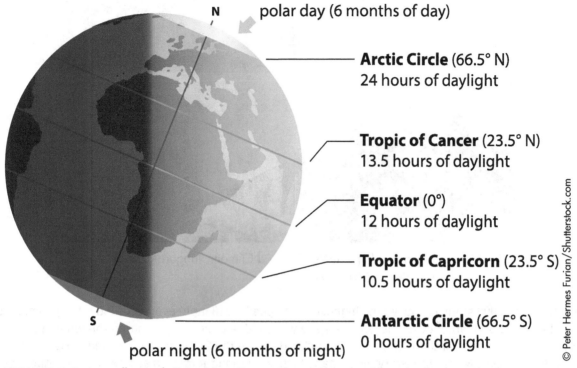

polar day (6 months of day)

Arctic Circle (66.5° N)
24 hours of daylight

Tropic of Cancer (23.5° N)
13.5 hours of daylight

Equator (0°)
12 hours of daylight

Tropic of Capricorn (23.5° S)
10.5 hours of daylight

Antarctic Circle (66.5° S)
0 hours of daylight

polar night (6 months of night)

© Peter Hermes Furian/Shutterstock.com

FIGURE 3.8 Annually, Earth Emits More Energy Than it Receives Above and Below 40° N&S.

Tornadoes, too, are seen by some observers as manifestations of global warming. It is interesting that some writers who carrying scientific credentials still use terms and expressions that would have been used by ancient writers in their assessment of divine retribution on wayward people. Geokolasis (earth punishes) was a term coined by Vann to express scientists' view that nature punishes people for wayward behaviors.[9] An example of geokolasis is found in a book authored by American geographer Wallace Akin. The book also provides an excellent historical account of the Tri-State Tornado of 1925. Nonetheless, notice how Akin uses "gluttony," one of the seven deadly sins, in his evaluation of the present state of the consequences of human-environmental Interactions:

> I, as do most scientists, believe that global warming is increasing the frequency of severe storms, and this is a situation little understood by the public and their elected officials, who tend to focus only on temperature changes. But it is more than just warming. Our gluttony for energy is loading the atmosphere with pollutants and increasing the greenhouse effect. This subject is far too complicated to discuss briefly, but be assured that we can expect more frequent and more violent acts of nature, more hurricanes, [and] more heavy deluges of rain with resultant floods and mud slides--and more frequent super tornadoes.[10]

Akin did not put a time frame on his prediction, so a person who endorses his view can, for example, point to the tornado of May 3, 1999 that ravaged the I-35 and I-44 corridors in Oklahoma as evidence of his

---

[9] Vann, *Forces of Nature*, 36–37.

[10] Akin, *Forgotten Storm*, 8.

© Eva Kali/Shutterstock.com

FIGURE 3.9 Volcanoes Impact Weather and Climate. Tungurahua Eruption in 2014, Ecuador.

prediction. The 1999 storm produced the highest wind speeds (318 mph) ever recorded.[11] Keep in mind that record keeping for weather events is less than 150 years old, and no one measured the wind speed of the massive 1925 Tri State Tornado. The earth is much older than the 74 years that elapsed between the 1925 and 1999 tornadoes. No doubt history hides many facts about long-forgotten and extreme weather events. Keep in mind that most of the one thousand tornadoes that touch down every year in the United States do so in open prairie areas where few people beyond storm chasers see them.

A third reason that complicates any quantification of human agency in climate change is volcanic activity; they emit greenhouse gases like CO_2, water vapor, and methane as well as particulate matter, so eruptions can increase temperatures through greenhouse gases or, conversely, cool climate through particulate cover that blocks out portions of the sun's rays. Indeed, particulate matter can create a cooling effect when they are suspended in the atmosphere, but when the particulates fall to earth, its greenhouse gases are put back into circulation where they can trap sensible heat. An example of how a volcanic eruption can impact weather was on full display in 1815 and 1816 when Mount Tambora, an active volcano situated on the island of Sumbawa in Indonesia, erupted. Particulate matter in the atmosphere blocked out enough sunlight to cool global temperatures over the next year. The effect of global cooling was so remarkable that geographers and historians refer to 1816 as the "Year without a summer."[12]

As mentioned above, climate change is a complex and highly debated social and political issue. Economic activity is certainly one of the complicating factors in the climate change debate, and unfortunately, it has political implications. For instance, supporters of free-market economics resist arguing for government regulations to lower carbon footprints, while those who lean more to an activist government model like socialism see curtailing economic activities through regulatory actions, especially those that involve carbon-containing products, as a necessary step toward cooling a planet made red hot by human activities. Since the acquisition of social and political power is a goal of socialists, it is hard to know if their arguments are examples of the exploitation of a possible problem or whether they are genuinely concerned about environmental conditions. Conversely, it is hard to know if free-market advocates who resist government regulations are making informed decisions or whether they care little for what humans might be doing to the

---

[11] Weather.com, "Most Extreme Wind Speed."

[12] Stothers, "Great Tambora Eruption," 1191–1198.

environment. Clearly, there are significant nonenvironmental incentives motivating proponents on both sides of the climate change debate. Finding objective analyses of the issue of climate change is a challenge. The topics of economics and political policies are further explored in Chapter 8.

While carbon-based fuels are obvious targets for activists, many nonfuel products are derived from petroleum. As a consequence, drilling for oil is not likely to be significantly reduced in the near future. Petroleum-derived products include aspirin, cortisone, plastics, tires, linoleum, denture adhesive, nail polish, basketballs, footballs, and dozens of other items. Consider that in a 42-gallon barrel of oil, only 19.2 gallons of gasoline is refined from it.[13] Finding a replacement resource from which to manufacture nonfuel items is a difficult, if not impossible, task. For better or worse, oil drilling and refining will be a part of life for years to come. The use of coal to refine nonfuel products is limited, so coal mining is much more likely to be regulated than drilling for petroleum.

Economics aside, the public in the twentieth century held wide-ranging views of climate change. One thing is for certain, atmospheric conditions have influenced societies' perceptions of life and the religious and political times in which they lived. It was not too long ago that Americans and Europeans feared the return of an ice age, not rising sea levels due to global warming.

The Ice Age Cometh

While global temperatures warmed from 1980 to 2010, an opposite situation marked the earth's climate during the decades from 1940 to 1979. There was a palpable fear that another ice age was developing. A story titled "The Ice Age Cometh" published on June 12, 1962 in the UK's *Guardian* reflects a view that was commonly held at the time. The piece shows a reasoned concern that another ice age was likely looming on the earth's temporal horizon. In answering a speculative question about the possibility of the recurrence of yet another ice age, the author wrote that "The answer . . . is most probably that there is. As evidence of climate in prehistory is accumulated, it becomes more and more evident that slow changes are continually taking place . . . On several occasions in the 1880s and 1890s, the Thames froze during winters

© Csaba Peterdi/Shutterstock.com

FIGURE 3.10 While often Frozen in the Little Ice Age, River Thames in England Rarely Freezes.

---

[13] Ranken Energy Corporation, "Products Made from Petroleum."

colder than any in recent memory. The river was still more thickly covered with ice in the seventeenth century; in 1680 there were eleven inches of ice on the Thames."[14] The author of that news article was correct about colder weather in the past. As mentioned briefly above, historians and geographers refer to the years between 1280 and 1850 as the Little Ice Age. The author was also correct in that there are continual changes, although not linear, in climate over time.

Before the Little Ice Age, there was a warm period that stretched from around 900 to 1250, which is known as the Medieval Warm Period or Medieval Climatic Anomaly.[15] It was during that warm period that the Vikings expanded across the North Atlantic and settled in Greenland. However, the harsh bleakness of the Little Ice Age pushed the Vikings to abandon their farms for presumably warmer climes.

Whereas during the Carboniferous Age, the earth's climate was balmy and warm with tropical vegetation growing as far north as Pennsylvania, continental glaciers were as much as two miles thick as far south as Kansas during the last Ice Age. Humans cannot live in glacier covered landscapes. Even during the Little Ice Age, shorter growing seasons and colder winters reduced the carrying capacity in rural farm areas. Some scientists and historians have argued that the Little Ice Age provided a major environmental push factor in mass migrations of peoples. For instance, in the 1600s, the English and Scottish governments, which were under one monarch, encouraged migration across the Irish Sea where vacant lands were granted to loyal planters who in turn leased lands to their loyal tenants. Whereas in the past, those tenants or peasants exchanged man-rent (labor) for farming privileges, the long-term leases in Ulster (nine counties in Northern Ireland) gave them the status of farmers. Within one hundred years of the Plantation of Ulster, the descendants of those pioneering farmers and merchants were themselves pioneers in colonial

FIGURE 3.11 Cherokee and Shawnee Moved South in this Region During the Little Ice Age.

---

[14] Guardian, "Ice Age Cometh."

[15] Henry and Hughes, *Medieval Warm Period*, 134; Mann, "Medieval Climatic Optimum," 514–516.

America. In America, many of them interacted with other climate-caused refugees: The Shawnee had moved into Ohio and western Pennsylvania while the Cherokee had relocated into southern Appalachia during the Little Ice Age. Both nations had migrated away from an area that extends from north of Lake Erie into what is today New York. Their mass migration enabled those who stayed behind to survive on the meager resources that the environment offered during the Little Ice Age.

It is no small wonder, then, why informed members of the public were concerned about changes in weather from 1940 to 1980. A return of an Ice Age would cause a rippling effect throughout societies around the world. Those ripples could include warfare or genocide as mass migrations extend over sovereign borders. People who lived east of Lake Erie were especially concerned about the return of an ice age. Lake effect snows are common in that part of the United States and Canada even during "normal years." In the 1970s, snows were deep, and they lasted long enough for subsequent snow-producing precipitation events to accumulate and form snow packs. Concern was raised because glaciers start off as snow packs. Indeed, glaciers are giant snow packs that have compressed so much that the snow is converted to ice. In Pittsburg, Pennsylvania, for instance, three of the five deepest snow packs in the city's recorded history occurred in the 1970s.[16] Two of the five deepest snow packs in Pittsburg were measured on March 14 and 15, 1993, months after the violent eruption of Mount Pinatubo in the Philippine islands. The United States Geologic Service (USGS) claims that the "dispersal . . . [of the volcanic] cloud around the world caused global temperatures to drop temporarily (1991 through 1993) by about 1°F (0.5°C)."[17] At least for a short time, the weather in 1993 resembled the conditions that prevailed in the decades between 1940 and 1980.

© Pierre Williot/Shutterstock.com

FIGURE 3.12 Icy Broderick Park Walkway, near Buffalo, New York, January 1, 2018.

---

[16] National Weather Service, "Snow Records for Pittsburg, Pennsylvania."

[17] The United States Geological Service, "Cataclysmic 1991 Eruption of Mount Pinatubo."

SUMMARY

The idea that climate change is a complex issue rests on a number of environmental factors as well as human activities, perceptions, and political and economic agendas. While science is making headway into understanding and even quantifying the relationship between solar activities and atmospheric temperatures, more research needs to be conducted to unravel the effect of solar energy absorption in tropical oceans. Through ocean currents, hurricanes, and tornadoes, nature diffuses sensible heat poleward. Also, more research is needed in the area of ground water mining and its effects on weather and climate. In the final analysis, climate change is far too complex to be reduced to simple political sound bites. Objective research on the nature and historicity of climate change needs to be conducted. By understanding the historicity of climate change, we will come much closer to understanding what it means for us relative to our response to it. In Chapter 4, we will dig deeper into how the ancient Celtic culture that stretched from what is today Turkey to Scotland and Ireland understood seasons and weather and climate. Many of their important dates throughout the year were based on earth–sun relationships that were tied to seasons and agricultural production. If you live in the West, you might be surprised to know how many of those dates were absorbed into Christianity and are still with us today.

CHAPTER 4

Ancient Celts' Relationship to Seasons and the Sun

INTRODUCTION

Many once thriving cultures of the ancient world no longer exist; most of them left no written record of their goings on. There is virtually no way of knowing how they perceived and interacted with the natural world. The Picts who lived in what would become Scotland, however, left some artifacts that have Latin wording. One of the best examples of this kind of linguistic artifact is a Latin inscription found on the Dupplin Cross, which was erected on the Dupplin Moor in 800AD.[1] The passage reads *Custantin Filius Fircus* (Constantine son of Fergus). Pictish-era crosses with Latin wording are common in Ireland and England, but they are rare in Scotland. Latin was certainly an ecclesiastical language used in the west of Europe, Great Britain, and Ireland in the 800s AD. However, the nonclergy members of the church most likely did not know how to read or speak Latin. Their native Pictish language is thought to be an insular Celtic language; however, it is now extinct and was most likely only a spoken tongue.

Older cultures like the many Paleolithic and Mesolithic villages that stretched along the shorelines of oceans were washed away as sea levels rose when the Pleistocene abruptly ended about 12,000 years ago. Their stone tools, animal hides, and arrow points have either decayed or are submerged beneath several feet of seafloor sediment. Other groups of Celtic people, however, left us many linguistic and material artifacts. Those artifacts reveal a great deal about how they understood earth–sun relationships, seasons, harvest and planting times, and worship practices. In this chapter, we will take a closer look at how the pagan Celtic peoples, who lived in lands that stretched from Galatia in modern Turkey to Ireland and Scotland, understood the fundamental relationship between the sun and earth as well as how that relationship impacted the ebb and flow of the seasons and life. Through those insights, we are able to glean aspects of climate that are somewhat familiar to us. Through their conversion to Christianity and intermingling with the Germanic peoples of northern Europe, many of their festivities surrounding earth–sun and seasonal dates have been handed down to us. It is remarkable how well they understood seasonal and solar cycles, for they are still accurate. Along the way, their festivities and dates morphed into familiar dates and holidays like Halloween, Christmas, and Candlemas, which is more commonly known today as Groundhog Day.

---

[1] The Dupplin Cross is on display at St. Serf's Church in Dunning, Scotland.

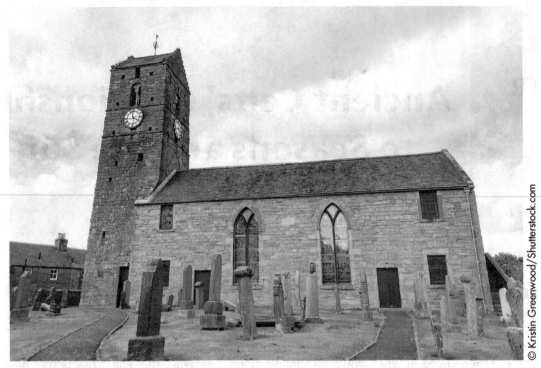

FIGURE 4.1 St. Serf's Church: The Home of the Dupplin Cross.

FIGURE 4.2 Pictish Cross.

FIGURE 4.3 Earth's 23.5° Tilt and Resulting Seasons.

Imagine a world in which there was no social security system, Amazon.com, Dollar General Store, or Walmart, to supply our needs and wants; we would be totally dependent on ourselves, close family, and maybe friends for survival. Major factors in sustaining life for those who lived during the medieval era were weather, land, and water resources, as well as the will and benevolence of deities. For those who lived between 30° N and 60° N, the northern middle latitudes, the earth's 23.5° tilt meant extreme changes in the seasons. The spring brought warm, balmy weather in which rainfall ushered in budding fruit and greenery. Summer was a time to nurture crops and watch fruits ripen to harvest. If all went well, bountiful harvests took place around the autumnal equinox that falls on or near September 21st each year. Still, we are left to wonder why the ancient Celts celebrated Samhain (Sowin), an end of summer harvest festival, on October 31st and their New Years on November 1st, the next day? Those dates are over a month past the fall equinox, and, if weather and climate were like those of our time, no doubt most of the harvested crops were already preserved or otherwise stored away. Another question arises from this observation: was the late date a reflection of longer-lasting seasons?

Why October 31st for Samhain?

Let us ponder a geographic theory to explain the dates. As most school kids know, the autumnal equinox is a day in which there are equal amounts of daylight and darkness at the equator; as mentioned earlier, it occurs on or near September 21st each year. You might recall that the same amount of daylight and darkness occurs on or around March 21st, but there is a big difference in air temperatures on those two dates. Because much of the sensible heat in the Troposphere, the layer of atmosphere in the northern middle latitudes which we live, is actually reradiated solar or longwave radiation that the earth and bodies of water had stored and converted from shortwave solar radiation, there is a time lag between the arrival and departure of seasons and solar angles associated with equinoxes and solstices. Water, for instance, has a higher specific heat (a rating for the amount of energy required to raise a unit of a given substance by one degree) than land. Large water bodies heat up and cool off much slower than land; therefore, land laying near a large water body will experience a moderating influence on its temperature range throughout the year. This means that even though the spring equinox has the same amount of daylight as the fall equinox, the weather is much warmer in late September than it is in late March, even in continental locations. Consider that Denver, Colorado, averages around 54°F during March, but its average high in September is 77°F.[2] That 23-degree difference in mean high temperatures is easy to explain: shortwave-solar radiation absorbed in soil, rocks, and water during the summer months is reradiated into the atmosphere as longwave radiation. That energy keeps early fall temperatures pleasant. In the second half of fall, enough of the terrestrial and aquatic energy reserves have been lost to space, and the decreasing temperatures we feel make the reality of an approaching winter all too obvious. Mornings are frosty, and at elevation, snow is beginning to blanket the rocky outcrops and slopes. To the ancients, October 31st and November 1st marked a sensible transition from the life supporting weather of late summer and early fall to the deathly challenges of the depths and darkness of winter. When I lived in Scotland, I played golf well past 11 p.m. in the middle of summer, but in the mid-winter, there was barely enough daylight hours to walk 18 holes, even if the weather permitted. Due to relatively warm ocean currents, playing golf was sometimes possible in winter months. As people who were deeply connected to the natural world and its rhythms, its highly probable that the date for Samhain and the Celtic New Year were tied to a sensible reality of the passage of the seasons of life into what they perceived to be the season of death. It does not mean that the dates for the seasons were much different than they are today.

Celtic Superstitions and Beliefs Involving Earth–Sun Relationships

With 40 being the new 20, it's hard for some modern folk to fully appreciate the fact that for most of human history few people lived beyond the age of 40. Outside of ancient Egypt, few cultures in the ancient world had embalming practices to keep corpses from swelling up and oozing gases and body fluids. The reality of death and the physical consequences of a loved one's decomposition were constant reminders of one's own mortality, especially in warmer months of the year. The Celts were simultaneously comforted and afraid of the transition to winter. Cold weather's threats to life and limb increased precipitously throughout the fall season, so it was a time of year in which beliefs about visits from deceased loved ones gave them some psychological comfort. The Celts specifically believed that the spirits of the dead were able to revisit the earth on the night before the New Year. Through their rituals, the people believed that Druids could receive predictions from the dead. The Celts came to believe that by dressing up in scary costumes, they could keep fearful spirits or ghosts at bay. To commemorate the event, Druids built large bonfires around which people would gather. They placed crops and animals in the fire to be consumed as sacrifices to Celtic deities. During those celebrations, people wore animal hides and the skulls of beasts. They also attempted to make predictions for each other. At the end of the evening, the sacred bonfire was used to light torches that the Celts took home to spark fires in their own hearths. They believed that the sacred flames would invite good luck during the dark and cold winter months.

By 43 AD, most of Celtic Europe, outside of Scotland, Ireland, and the Celtic fringes of Great Britain, were under Roman Rule. Some scholars contend that the Roman festival of *Feralia*, which was held on a day in late October to commemorate the dead, was absorbed into Samhain. They also believe that a festival to honor Pomona, the Roman goddess of fruit and trees, influenced the tradition of bobbing for apples.

---

[2] U.S. Climate Data.

All Saints' and Reformation Days

The Roman Emperor Hadrian completed work on the Pantheon ca. 124 AD. The building has been in continual use since its construction. When Christianity was embraced by the Roman Empire in the fourth century, the Pantheon played a vital role in expressing the empire's new faith. On May 13, 609 AD, Pope Boniface IV dedicated the Pantheon to the honor of Christian martyrs. A century later, Pope Gregory III, who was born in Syria and the last Pope born outside of Europe until Francis was elected in 2013, expanded the festival to include all saints as well as martyrs. He also changed the date of observance from May 13 to November 1st, the day of the Celtic New Year. In 1000 AD, about the time the Vikings landed in North America, the Roman Catholic Church named November 2nd All Souls' Day. It was a day to honor the dead. It is believed by many that the church was doing its best to supplant the Celtic Festival of the dead. As is shown shortly, the church had made a similar move to supplant a Celtic solstice festival by setting December 25th as a day to observe Christmas.

© Julian Schlaen/Shutterstock.com

FIGURE 4.4 Roman Pantheon Used by Pagans and Christians.

FIGURE 4.5 L'Anse Aux Meadows, Viking Settlement, Canada.

FIGURE 4.6 The Catholic Faithful Celebrate All Saints' Day in the Philippines.

All Saints' Day was celebrated in a similar fashion as the Celtic Festival of Samhain. The festivities included lighting large bonfires, dressing up as angels, devils, and saints, as well as taking part in parades. All Saints' Day was also called All-hallows, which in the Middle English was Alholowmesse. The night before All-Hallows, the same day as the Celtic festival Samhain, was called All Hallows Eve. Eventually All Hallows Eve became Halloween.

Christianity's involvement with October 31st, as an important date in salvation history, did not stop with the Roman Catholics. According to the Protestant Reformer Philip Melanchthon, who was a collaborator of Martin Luther, the Protestant Reformation began on October 31, 1517 when Luther nailed his 95 Theses to the door of the University of Wittenberg. In some learned Protestant circles, October 31st is better known as Reformation Day, not Halloween.[3]

Although New England with its fall colors conjure up images of Halloween, English Puritans actually kept celebrations of any festivals to a minimum. As a Catholic-friendly colony, Maryland was more receptive to recognizing the day. The southern backcountry of Appalachia was settled by Irish, Welsh, and English settlers from the Celtic fringes of Cornwall and the border country; they, too, were eager to keep their traditions alive.[4] Keep in mind that the Romans never conquered Ireland or Scotland, and their imprint on Wales and Cornwall was limited. The border country was so volatile that the Roman Emperor Hadrian built a wall across the isthmus of Great Britain, effectively writing off any prospects of pacifying the Picts who were one of the cultural parents of the Scots.

FIGURE 4.7 Martin Luther Holding His Translation of the Bible in Wittenberg, Germany.

[3] Martin Brecht, *Martin Luther*, 200–201.

[4] Vann, *South's Celtic Heritage*, 139–161.

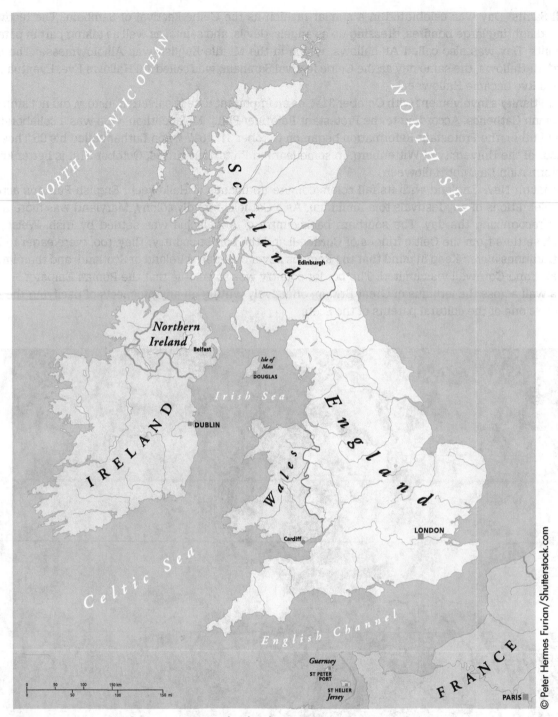

FIGURE 4.8 Map of Great Britain and Ireland.

Just as Halloween is a blend of Celtic pagan and Christian influences, it is no small wonder why the southern uplands of the United States still celebrate stories about ghosts, goblins, and things that go bump in the night. As a friend reminded me, the noted author C.S. Lewis was a Protestant born in Belfast, Northern Ireland; he was accused of being soft on paganism. Even for those who are not pagan believers, ghostly Celtic stories are entertaining.[5]

_____

[5] Ibid.

Winter Solstice

Christmas is a favorite time of the year for many people, even if every major automobile manufacturer wants us to think that gifting people with cars is commonplace. Car commercials aside, Christmas is a global event, but because of its origin, dare I say the reason for the season, which centers on the birth of Jesus of Nazareth, the holiday serves a mostly commercial function in the non-Christian world. As is shown in this section, Jesus's birth date was superimposed on a Celtic date that was used for festivities celebrating the winter solstice. Still, even in the historically Christian West, there are five persevering myths about Christmas's origins and history: Myth 1: Jesus's birth year was not the year 1 AD; by using dates associated with known historical figures like King Herod who was alive at the time of Jesus's birth (Matthew 2:16) but died in 4 BC, most theologians believe Jesus was born between 6 BC and 4 BC.[6] Myth 2: December 25th was not likely the day of Jesus's birth. There is a 1 in 365 chance that Jesus was born on December 25th.[7] Myth 3: The person of Santa Claus or Saint Nicholas is a myth. Nicholas was an actual Christian bishop who lived in the 4th century AD. Myth 4: St. Nicholas was most likely not a ruddy complexioned man from Germany. He was a Greek living in a village named Myra in what is today the Mediterranean coast of Turkey. Myth 5: St. Nicholas traveled about in a sleigh. He would have used wagons and boats. He did not employ reindeer, or caribou as they are known in North America. Reindeer lived about 2,000 miles further north than St. Nicholas. It is doubtful that he had even heard of a reindeer, let alone seen one. The truth of Christmas is much more fascinating than the myths surrounding it.

Two of the most prominent symbols that we associate with Christmas (Santa Claus and the Christmas tree) have a decidedly Christian and Germanic influence. Even sacred holidays have a historical geography. Therefore, where did the starry night of Christmas, Santa Claus, and the Christmas tree originate?

First Global Holiday

Answering how Christmas traditions like Santa and the tree spread around the world is easy: as Christians and missionaries migrated around the world, they took their holiday traditions with them; because gifts and sharing were associated with those traditions, commercial activity was not far behind. Because of the festivities and commercial implications of the Christmas holiday, many non-Christian societies delight in festive trees and lights as well as Christmas malls.

FIGURE 4.9 Christmas in Chengdu, China.

---

[6] Filmer, "Chronology of the Reign," 283–298; Steinmann, *From Abraham to Paul*, 235–238.

[7] The Gospels of Luke and Matthew, which are the only locations in which Nativity scenes are described, do not mention a date or season for his birth.

However, Christmas carols with religious themes are not so popular in many parts of the world, but to Christians, spiritual hymns sung at Christmas are sacred and full of geographic wonder. Consider some of the lyrics to "Silent Night"; it was written in 1818 by Franz Gruber and Joseph Mohr. Those fellows lived in a small village near Salzburg, Austria.[8] In recognition of its widespread popularity, UNESCO named "Silent Night" an immaterial artifact of cultural heritage in 2011. Even though no one knows for certain the day in which Jesus greeted the earth as a baby dressed in swaddling clothes, culture has nurtured the idea that he was born on a crisp winter night.

Winter Sky On the Night of Christ's Birth

Being from the hilly country of East Tennessee, which looks a lot like the land around Salzburg, Austria, and the birthplace of "Silent Night," it is easy for me to attach similar geographic images to the scenes depicted in "Silent Night." Perhaps it is the same for you. Where I come from, still, starry skies are associated with crisp autumn or winter nights. In my imagined world of "Silent Night," the shepherds are huddled together for warmth as their sheep browse on frost-covered grass. The lyrics of "Silent Night," to me, create an image of a completely peaceful, pastoral scene far removed from the hustle and bustle of first-century urban life. Other popular carols also feature peaceful star-lit nights: "Away in the Manger," "O Holy Night," and "We Three Kings."

As mentioned above, however, the telling of the nativity is only discussed in Mathew and Luke. The imagery in "Silent Night," "O Holy Night," "Away in the Manger," and "We Three Kings" is a blend of the ways in which Jesus's birth is depicted in those two gospels plus some seasonal interpretation.

© Zvonimir Atletic/Shutterstock.com

FIGURE 4.10 Angels Informing Shepherds of Jesus's Birth.

---

[8] Daley, "It's the Bicentennial of 'Silent Night.'"

The night of the nativity seems still and starlit, but Scripture only mentions one bright star; for residents in the northern middle latitudes, which is where most people live, still nights are associated with high-pressure centers in the winter; it is easy for us to accept that the nativity occurred in winter. However, we really do not know at what time of year it happened.

Consequently, you might ask, "Why do we celebrate that date"?

The early church wanted to set a date for "Mass on Christ's day" (Christmas). The church leaders were especially interested in supplanting an existing pagan holiday known as Saturnalia; the date, December 25th, occurred just after the winter solstice when a brief hint of increasing daylight suggested the return of spring, which they believed signified a rebirth of life through Christ. Symbolically, Scripture is in harmony with that logic, for in John 8:12, Jesus refers to himself as "the light of the world." December 25th, as the official date for Christmas, first appears in *Chronographiai*, a five-volume treatise on history written in 221 by a Christian historian named Sextus Julius Africanus.

We can be confident that the Christian significance for Christmas originated somewhere between 6 BC and 4 BC in a small Judean village named Bethlehem, but its place on the calendar rests with the ancient Celts who based many of their important dates on earth–sun relationships and the passage of seasons.

While the early church was shaped by such thinkers as Paul, Peter, Augustine of Hippo, and Justin Martyr, among others, it is interesting how a lowly bishop working in Myra, a small village on the north shore of the Mediterranean Sea, come to be associated with the birth and life of Christ? It was St. Nicholas's charitable deeds that led to myths of Santa Claus.

Groundhog Day (Woodchuck Day)

Because of the name of this chapter, reading that Groundhog Day is of Celtic origin will not come as a surprise to you. As with the annual date for Christmas, its date was set on a Celtic holiday known as Alban

FIGURE 4.11 Does St. Nicholas's Real Home Town of Demre, Turkey Look Like the North Pole.

Arthan (also known as Yule) that involved earth–sun relationships. Alban Arthan celebrations were led by priestly Celtic Druids on the winter solstice; they saw that date as one of the most important days of the year. In Wales, Alban Arthan meant "light of winter." In England, it meant the "light of Arthur." The Roman historian, Pliny the Elder (23 AD–79 AD), who died long before the birth of the Arthurian legend, recorded some of the practices that the Druids performed on Alban Arthan. The elder historian described a practice of gathering the parasitic mistletoes from Oak trees to hang over doorways. Because Mistletoe seeds contained a sticky white paste that resembles semen, the Druids believed that the mistletoe was a symbol of male fertility.[9] In time, Celtic boys and young men saw the occasion as a license to kiss girls who walked under a hanging mistletoe.

As with the Celtic harvest festival known as Samhain (Sowin) or All Hallows' Eve, the early church chose a date for Christ's mass or Christmas to coincide with Alban Arthan and the Roman Festival Saturnalia. Church leaders chose December 25th, which falls just past the winter solstice, as the day to celebrate Christ's birthday.

As a halfway point on the calendar between the winter solstice and the vernal or spring equinox, the Celts observed *Imbolc* or the rebirth of spring on February 2nd. Interestingly, in modern times, we recognize the first day of spring as falling on the equinox (equal day and equal night). The spring equinox falls on or near March 21st. As Christianity spread throughout Europe, *Imbolc* evolved into Candlemas, a festive event that celebrated the presentation of Jesus at the Jewish temple in Jerusalem. Luke 2:22 to 23 describes how Joseph and Mary presented themselves and baby Jesus at the temple for their purification and for Jesus's designation as "holy to the Lord." Since Jesus would have been 40 days old, and it was

FIGURE 4.12 Names of Celtic Festivals.

---

[9] Dunn, "Mistletoe."

33 days since his circumcision, the early church chose February 2nd (the date for Imbolc) as Candlemas.[10] As with the date for Christmas, however, no one really knows the day of the year in which Jesus was born, let alone the day in which he was presented to the rabbis at Jerusalem's holy temple.

Since Jesus's birth signified new life for those who believed in him as their savior, the date for his presentation came to be associated with the Celtic idea of the birth of spring. In certain parts of Europe, many Christians came to consider that if Candlemas was a sunny day, there would be 40 more days of cold and snow. If it was cloudy on Candlemas, they reckoned that an early departure of winter weather was probable. As Christianity and its traditions, like Halloween (Samhain), Christmas and Santa Clause, spread into Germania, the Germans put their own mark on Candlemas.[11] They came to believe that if hedgehogs or other small animals saw their shadows on February 2nd, it was truly a sunny day. Only then could they expect six more weeks of cold and snowy weather.

Eighteenth- and nineteenth-century German immigrants brought their holiday traditions with them to Pennsylvania. With more groundhogs calling the Keystone state home than hedgehogs, German settlers recognized the groundhog as nature's meteorologist. The groundhog is also called a woodchuck, and because they whistle when they are frightened or trying to mate, they are sometimes called whistle pigs. Weighing only 12 to 15 pounds fully grown, the groundhog is a rather small animal to be likened to a pig.[12] Nonetheless, the groundhog's hibernation cycle makes it an excellent, though not accurate, candidate to serve as nature's traditional meteorologist. They go into hibernation in late fall. During that time, they lose 30 percent of their body fat. Their body temperature lowers significantly, and their heart rate drops from 80 beats a minute to about five. In early February, male groundhogs emerge from hibernation to look for a mate, not to predict the weather; that is when they are likely to be heard whistling. They do not stay out for long, however. They return to hibernate some more, but they come out for good in March.

© Brian E Kushner/Shutterstock.com

FIGURE 4.13 Groundhog Emerging from Hibernation.

---

[10] Knecht, *A Practical Commentary*, 410.

[11] History.com, "Groundhog Day."

[12] The History Channel, "First Groundhog Day."

In 1887, a Pennsylvania newspaper editor, who belonged to a club called the Punxsutawney Groundhog Hunters, encouraged his club to name their mascot, a wee whistle pig named Phil, to serve as America's official weather prognosticator.[13] Since groundhogs only live 6 to 8 years, there have been many woodchucks playing the role of Phil the forecaster. Not to be outdone, other cities have joined in the fun. New York has Staten Island Chuck; Canada has Shubenacadie Sam, and Alabama has Birmingham Bill.[14] Still, the grandfather of all Groundhog Day celebrations is held at Gobbler's Knob in Punxsutawney, Pennsylvania, every February 2nd.

With tens of thousands of people taking part in annual Groundhog Day celebrations, one might wonder about Phil's accuracy. "A team . . . at *The Washington Post* actually did the math. They calculated the average daily temperatures during the six weeks after Groundhog Day for the past 30 years";[15] They compared temperatures during the years in which Phil saw his shadow to the years in which he did not. "In the Post's analysis, weather records suggest that Phil was right more often than not in some cities, but the results basically come down to chance. Temperatures do not vary uniformly across the country, so Phil's predictions are bound to be right somewhere."[16] Take for instance, the years in which Phil saw his shadow, Oklahoma City was on average 8.5°F (4.7°C) cooler than during years when he did not see it.[17]

FIGURE 4.14 Entrance to Gobbler's Knob, Punxsutawney, Pennsylvania.

---

[13] Ibid.

[14] Ibid.

[15] Lewis, "Groundhog Day."

[16] Ibid.

[17] Ibid.

FIGURE 4.15 St. Petersburg, Florida.

However, in other cities the admired rodent's predictions were not remotely close. "For example, in St. Petersburg, Florida, years when Phil saw his shadow were on average 13.9 F (7.7 C) warmer than non-shadow years."[18]

SUMMARY

The ancient Celts were certainly interested in earth–sun relationships. They based most of the dates for important festivals on where the sun was located in the sky. They understood that the solstice was a day in which the shortest amount of daylight was seen. Conversely, the day in which the most sunlight was seen occurred on the summer solstice. They also knew that March 21st was an equinox, a day in which there is as much sunlight as there is darkness. No doubt they detected increasing sunlight after the solstice and made observations that some plants produced buds by early February. They reckoned that the halfway point between the winter solstice and the vernal or spring equinox was a date that signified the return of spring. As Christianity was embraced by the Celtic peoples of Europe and Asia Minor, the church absorbed many of the important dates from the pagan calendar and attached to them unknown yet significant dates for events found in the Bible. It is important to know that the Book of Galatians in the New Testament was written to a Celtic Tribe of the same name. Because the Bible was written over an extended period of time, there are many biblical narratives that depict weather and climate conditions. It is a comparison of modern weather conditions in the Biblical lands to those narratives that we now turn our attention.

---

[18] Ibid.

Weather and Climate in the Modern Levant

INTRODUCTION

Even though Abrahamic religions ushered in an era in which their adherents jettisoned nature gods in favor of a personal deity who transcended all realms of life, control of natural forces was nevertheless retained by the God of Abraham. The authors of the texts associated with Judaism and Christianity, which were written over a period of time that spanned 600 to 1800 years, depending on whose point of view one takes, show that ancient and medieval people had a solid appreciation or reverential fear of the forces of nature. While the great flood event that all Abrahamic faiths embrace has made it into popular culture, a close examination of Scripture shows many other weather and climatic conditions. Imagery in biblical narratives, therefore, provides tantalizing glimpses into weather and climatic conditions at the time each respective author penned his book. These climatic and meteorological events show temporal patterns or epochs. Geotheological imagery shows that climate change is evidenced in Scripture, and because there are temporal patterns among meteorological events, it is possible to show that the Old Testament was written over at least a 1400-year span of time. Distinct weather patterns in the Bible are aligned with the Book of Job, the Pentateuch, most of the history books, and the New Testament. The wisdom books (Job, Proverbs, Song of Solomon, Ecclesiastes, Psalms) are spread out over more than one epoch, so there are a variety of weather conditions seen in them.

Most of the Old Testament authors were not well-travelled, so their environmental perceptions and geographic knowledge (geosophy) were limited to conditions with which they were familiar through personal experience or from others who had talked about other places. By examining scriptural narratives in light of paleoclimatological patterns and in contrast to modern climate conditions, it is possible to delineate more complete "past geographies" of the Levant.[1] In the pages that lie ahead, geotheological concepts developed by J. K. Wright[2] are infused with an ethnomethodological paradigm described by sociologist Harold Garfinkel to frame narratives of biblical weather and climate. The results of the analysis discussed in this and in Chapter 6 can help us further establish general time periods in which books of the Bible were written, which as is shown shortly, is a debated topic. This approach to understanding past climates is supported by an observation made by Kaylene Fleming who argues that delineating the manner in which past societies perceived environments and their meanings for archaeologists can be "considered [by answering] two questions posed by Kenneth H. Craik . . . 1. How do newcomers and natives differ in their

---

[1] H. C. Darby highlighted three distinct approaches to historical geography, including "past geographies," "history behind geography," and "geography behind history." See Philo, "History, Geography, and the Still Greater Mystery of Historical Geography," 252–282.

[2] Wright. "Terrae Incognitae," 1–15; "Notes on Early American Geopiety."

perceptions of a landscape? 2. How do the backgrounds and prior personal environment of each observer affect perceptions of a specific setting?"[3] Because weather and climate are active agents in any environmental setting, it stands to reason that Fleming's observation about a "specific setting" should include atmospheric conditions.

The purpose of this chapter, therefore, is to provide a foundation of knowledge about biblical dating techniques, framing geotheological concepts and their application to weather and climate imagery in the Bible, as well as the mechanics and contemporary weather and climate conditions in the modern Levant. In Chapter 6, this knowledge will be contrasted with narratives written during various biblical time periods. The evidence presented in this and in Chapter 6, therefore, suggests that some books of the Old Testament were written early in the time periods that are debated by biblical scholars.[4] The Book of Job may well have been penned earlier than most of the ancient dates suggested for it, although a few sources date its writing to before 2000 BC, which is not substantiated by paleoclimatological studies. Indeed, the Institute for Creation Research argues that Job was written during the time of Adam and Noah:

> Except for the first eleven chapters of Genesis, which almost certainly were originally written by Adam, Noah, the sons of Noah, and Terah, then eventually edited by Moses (compare with "Introduction" to Genesis), the book of Job is probably the oldest book in the Bible. It contains more references to Creation, the Flood and other primeval events than any book of the Bible except Genesis, and provides more insight into the age-long conflict between God and Satan than almost any other book. Remarkably, it also seems to contain more modern scientific insights than any other book of the Bible.[5]

As is shown in the next section, there is tremendous disagreement over dates of authorship for the Old Testament. Weather and climate patterns in Old Testament books do not support the recent authorship position, but their use as a means to date Scripture has not yet matured.

Using Weather and Climate to Help Date Books in the Bible

Dating books in the Bible has heretofore been based on references to cultural items (bronze vs. iron), historical events that can be dated, and which documented kingdom or ruler was mentioned in the present or past tense. Among religionists who study the dates of the books in the Bible, there are, as is shown shortly, at least two major schools of thought on when the texts were penned; however, there is some general agreement on the chronology of the books. Because there are no references to Mosaic Law or the exodus, for instance, Bible scholars generally agree that the Book of Job is the oldest biblical text. Depending on the scholar, Old Testament writing, with the exception of the Book of Job, can be placed somewhere after 1500 BC[6] to as late as the third century BC.[7] Genesis 4:22 tells us that Tubal-Cain forged tools out of bronze and iron. The transition from bronze to iron tool-making among the Hittites in the Levant occurred around 1400 BC.[8] Some of the Old Testament books were written during or just after the Babylonian exile, so their dates cannot be older than the sixth century BC. I Chronicles was probably written ca. 400 BC due to genealogical evidence. Dates for the New Testament range from mid-first century through early in the second century.[9] In general, therefore, there appear to be two major camps in the literary world of biblical scholars: older

---

[3] Fleming, "Cultural Landscape," 112–117.

[4] In this paper, *The* King James and the *New King James version of* the Bible were used. These versions of the bible provide more accurate word for word translations. This is especially important because, for example, the New International Version (NIV) uses "Dead Sea" instead of the Salt Sea, which was more the custom in the biblical era.

[5] "Institute for Creation Research, "Discover and Defend."

[6] See, for instance, Brindle et al., *The King James Study Bible*.

[7] Rogerson, "The History of Tradition," 8.

[8] Clark, *The Penguin Dictionary*, 213.

[9] Kim, "Letters of John," 250.

text advocates[10] and the later or recent authorship proponents.[11] As noted above and more fully explained later, weather and climate imagery in the books of the Bible suggest that the Book of Job, the Patriarchs, and the Pentateuch favor an older authorship (A more detailed analysis of weather patterns is given in Chapter 6). For a thorough analysis of the challenges and methods used by scholars to date books of the Bible, see *The New Bible Dictionary Second Edition*.[12] The use of weather and climate to date Scripture is scarce among biblical scholars. This is perhaps due to a lack of interest or background in physics and systematic geography among Bible scholars. It could also be because of the politicized nature of current discourses in climate change. Bible scholars already work in a field in which there are diverging and deeply held beliefs about the nature of God, the environment, the self, and nation, so adding the complex topic of climate change to the mix of variables used to place books of Scripture into temporal categories may well be a "bridge too far." In an opposite sort of way, many scientists who study weather and climate may shy away from Scripture as an objective source for evidence in analyzing changing climate.

Framing a Geotheological Approach to Biblical Climatology

Since geoscientists regularly map and measure observable and existing cultural and natural landscape features, most of their studies are set in a modern context because of the limited availability of scientific data from past times. The mapping of religious behaviors has been around since 1973 and the publication of Wilbur Zelinsky's *Cultural Geography of the United States*; cultural, physical, and social geographers, nevertheless, generally avoid delving into the thought processes of religious leaders and their followers. This avoidance exists because most geoscientists have regarded such activity as falling within an area of study relegated to philosophers and religionists, as well as historians who study past institutions and social movements; moreover, many believe that religions are inherently biased due to the limited set of believes on which they are based. However, a relatively small number of scholars have argued that an ethnomethodological approach to religious thinking helps inform an understanding of the anthropogenic or cultural landscape as well as the direction and intensity of movement of people across space.[13] Indeed, the existence of texts and archeological evidence from ancient times has enabled some scholars to develop a subfield that considers the interfaces between the imagined worlds of religious leaders and the ways in which their adherents have perceived and discussed human–environmental interactions, which is the essence of Garfinkel's conception of ethnomethodology.[14] Please note that the use of "imagined communities" and "imagined worlds" are not used to insinuate that these concepts have no semblance to real groups of people and places. Consider what it would be like to visit the North Pole or Antarctica. Unless you are one of a fortunate few to have visited those places, you nevertheless have an image of what those places would be like if you were to pay a visit. Our imagined communities and worlds are based on what we know about people and places that we may have never met or otherwise experienced firsthand.

In the realm of places, nature, and nations, much is owed to the foundational lexicon of John K. Wright and his notion of geotheology. A few geographers, since Wright's seminal work on geotheology in the 1940s, have used terms associated with his lexicon. Wright's geotheology, which ties together place and the worship of the divine, is similar to ethnomethodology, although understandably place is stressed more in geotheology. Still, few geoscientists attempt to enter the minds of long-dead people whose actions were influenced by deeply held beliefs.

---

[10] Douglas et al., *New Bible Dictionary*, 188–193.

[11] Rogerson, *Eerdmans Commentary*; Kugler and Hartin, *Introduction to the Bible*, 193.

[12] Douglas et al., *New Bible Dictionary*.

[13] Wright, "Terra Incognita," 33–35; Zakai, *Exile and Kingdom*; Mullan, *Scottish Puritanism*. In his book, Mullan analyzes social networks and world views of the leaders of an imagined community of ministers who led an untold number of their followers across the Irish Sea.

[14] Garfinkel, "The Origins of the Term Ethnomethodology," 15–18.

© Liz Glasco/Shutterstock.com

FIGURE 5.1 Mount of Olives Jerusalem Cemetery.

Because the sacred texts of Judaism and Christianity were written over an approximately 1600-year span of time, the way in which weather and climate are used by the various authors tells us much about how they saw natural conditions in the Fertile Crescent. Biblical imagery provides the basis for reconstructing weather and climate conditions of the pre-Greek and Roman eras. Before presenting an analysis of scriptural passages in Chapter 6, it is helpful to cast some light on the prevailing weather patterns that characterize eastern Mediterranean climate that is found throughout a large portion of the Abrahamic lands or the Levant.

Modern Era Climate and Weather: A Basis for Comparison

Most of the books that make up the Bible were written by people who presumably had a restricted knowledge of the world outside of the Fertile Crescent. Aside from the Apostles Paul and Peter, a few of the New Testament authors ventured beyond the Levant; however, New Testament writers may have learned about the climatic conditions of distant places by interacting with Romano–Hellenistic travelers. It is not well documented that many of the Old Testament writers travelled further north than what is today southern Syria, but Scripture points out that some of the Old Testament authors lived in Mesopotamia (modern Iraq), Persia (modern Iran), and Egypt during their generations' respective exiles. Writers of sacred texts were most likely describing familiar and even aberrant atmospheric conditions with which they were familiar. It is, therefore, important to contrast scriptural imagery with recent weather patterns for Jerusalem, Amman, Riyadh, Baghdad, and Cairo.

FIGURE 5.2 Tigris River Flowing through Baghdad, Iraq.

FIGURE 5.3 Modern City of Riyadh, Saudi Arabia.

FIGURE 5.4 Nile River Flowing through Cairo, Egypt.

FIGURE 5.5 King Abdullah Mosque; Amman, Jordan.

FIGURE 5.6 Panoramic View of the Old City, Jerusalem.

Jerusalem, which sits at an elevation of 757 meters (2,483 feet), has average January high and low temperatures of 11.89°C and 4.11°C (53.4°F and 39.4°F); one would not expect to find biblical accounts of snowy or icy conditions, unless they were seen as unusual.[15] Snow flurries occur in Jerusalem twice a year and a measurable snow fall happens about every three to four years.[16] Snow does infrequently fall elsewhere in the Near East, but any accumulation is short-lived. However, there is enough snow on Mount Hermon (7,336 feet or 2,236 meters), the highest point in the Golan Heights in Northern Israel, to support Israel's lone ski resort.[17] At 3,902 feet, Mount Meron is the second tallest mountain in Israel.

Amman, Jordan, which is not far from the ancient land of Edom and the likely location for the setting for the Book of Job, the oldest of the Abrahamic texts, has nearly the same average January temperatures as Jerusalem.[18] Riyadh, which is centrally located on the Saudi Peninsula, is a good bit warmer in the depths of January; its average temperatures range from 8.22°C to 20.0°C (46.8°F–68°F).[19] January is slightly cooler in Cairo, Egypt; temperatures range from an average low of 8.89°C (48°F) to average highs around 18.78°C (65.8°F).[20] Throughout the region, September through January are the wettest months with annual rainfall totals ranging from 2.54 cm (1.0 inches) to 58.9 cm (23.2 inches). January is Baghdad's coldest month with daily averages ranging from 5°C (41°F) to 10°C (50°F).[21] In many stretches of land situated between Cairo and Baghdad, there is no rain in the driest late spring and summer months, which is when tornadoes are most likely to form in the northern hemisphere.

© makearenko7/Shutterstock.com

FIGURE 5.7 Mount Hermon, Israel's Only Ski Resort.

---

[15] World Meteorological Organization (WMO), cited in *The World Almanac and Book of Facts*, 320.

[16] World Atlas. "Does it Snow in Jerusalem?"

[17] Tourist Israel: The Guide. "Mount Hermon Ski Resort."

[18] The Weather Channel, January 2019 data for Amman, Jordan.

[19] World Meteorological Association, 320.

[20] Ibid.

[21] Weather Online, "Iraq."

FIGURE 5.8 The Meron Mountains in Galilee.

FIGURE 5.9 A Heavily Irrigated Date Palm Grove in Israel.

Mechanics and Geography of Tornadoes, Whirlwinds, and Hail

To more fully appreciate the weather analysis presented in Chapter 6, it is helpful to explore the basic mechanics of weather events that are caused by cold air masses and their collisions with warmer air. It is helpful to recall that weather and life mostly occur in the troposphere, the lowest layer of the atmosphere. As is shown in Figure 5.10, the troposphere gives way to the stratosphere at about seven miles near the equator and at lower altitudes toward the poles. There is a distinct change in air temperatures that divides the troposphere and stratosphere. With warmer air found in the stratosphere, a temperature inversion called the tropopause forms on top of the troposphere; as its name suggests, it causes a pause in the rising of air masses. Because of this ceiling, weather and winds that affect life in the biosphere occur between the earth's surface and the tropopause. From a commercial air plane flying above 35,000 feet, it is easy to see that most of the weather is occurring below the plane (see Figure 5.11).

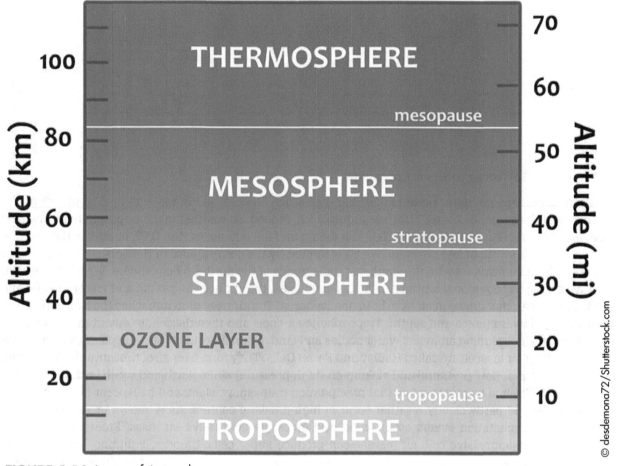

FIGURE 5.10 Layers of Atmosphere.

© JoeyPhoto/Shutterstock.com

FIGURE 5.11 Weather under the Tropopause.

There is perhaps no better place to start our discussion on cold weather events than to introduce the Intertropical Convergence Zone (ITCZ) (see Figure 5.12). Indeed, all weather in the troposphere is driven by incoming solar radiation or insolation. Near the spring and fall equinoxes, the ITCZ straddles the equator. It is effectively a swath of low pressure, which is formed by the convergence of the northeast trade winds in the southern hemisphere with the southeast trade winds of the northern hemisphere. With intense solar energy striking the equatorial latitudes near the equinoxes, the surfaces of ground and water heat up and produce rising air that draws in air from both hemispheres. The air rises until it reaches the tropopause and then shifts poleward (north and south). Tropospheric air cools and then descends toward the earth's surface. When the descending air meets water bodies and land, it spreads out toward the poles too. These rising and sinking air masses are called Hadley and Ferrel Cells. They form over all of the latitudes. The entire system of rising air (low pressure) and sinking air (high pressure) shifts north and south with the seasons (see Figure 5.12). It is helpful to know that precipitation (rain, snow, sleet, and hail) occur in low-pressure centers. Clear and partly cloudy weather occur in high-pressure centers. As is shown in Figure 5.13, the most volatile precipitation events occur when cold air is the aggressive air mass. Frontal precipitation events involving aggressive cold air masses can produce super cells, thunder, lightning, torrential rain, sleet, and hail as well as tornadoes that require upper level wind shears created by the jet stream. Heat generated precipitation events like those that occur in summer afternoons can cause hail, thunder, lightning, and torrential downpours. Heat generated events generally have more bluster than destruction and are generally short-lived, lasting less than an hour in duration.

A change in the earth's tilt would certainly impact the location of the ITCZ. If the earth's tilt would increase from 23.5° to 24.5°, temperatures throughout the year would be more extreme. Conversely, if the angle was reduced to a low of 22.1°, temperature variations would be milder.[22] According to the Milan-

---

[22] Graham, "Milutin Milankovitch."

kovitch Theory, a model developed by a Serbian climatologist and mathematician named Milutin Milankovitch (1879–1958), ice ages have resulted from simultaneous changes in the earth's tilt, shape of the earth's plane of orbit, and precession of the earth's perihelion and aphelion. We have already considered the effect that changing the earth's tilt would have on climate, so let us now take a look at climate changes that would occur if the shape of the orbital path were to change. Currently, the earth is at its closest point to the sun (perihelion) on January 5th, and it is at its furthest point on the orbital path on July 4th. When the plane of orbit is highly elliptical, there is a 20 to 30 percent increase in the amount of insolation entering the biosphere on the perihelion as there is on the aphelion. The last factor involved in Milankovitch Theory is the precession. In this part of the model, the earth's equinoxes move along the plain of orbit. Consider how weather and climate would change if the current perihelion, which occurs near the winter solstice in the northern hemisphere, were to occur on July 4th near the summer solstice. Insolation in the northern hemisphere would increase by 20 to 30 percent, so temperatures would be unbearable in many places north of the equator during July and August. According to the theory, which has been supported by deep sea sediment studies, thousands of years pass between simultaneous occurrences of these earth–sun factors.[23]

These Milankovitch factors are important to consider because many of the cold-weather events we see in Scripture are quite dissimilar to the atmospheric conditions we see today.

SEASONAL MIGRATION
OF THE INTERTROPICAL CONVERGENCE ZONE (ITCZ)

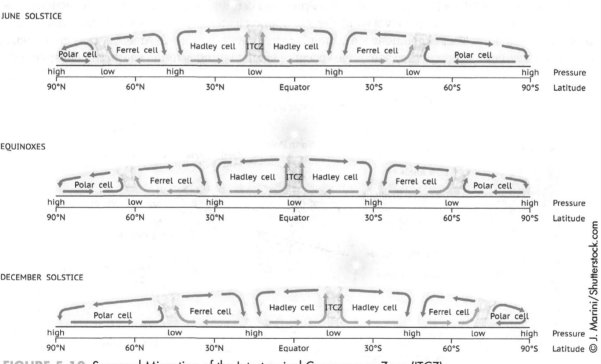

FIGURE 5.12 Seasonal Migration of the Intertropical Convergence Zone (ITCZ).

---

[23] Hayes et al., "Variations in the Earth's Orbit," 1121–1132.

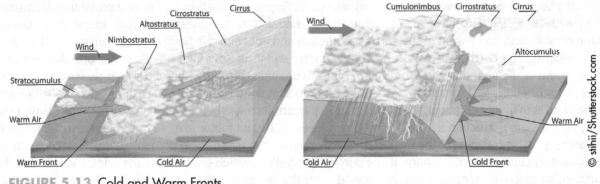

FIGURE 5.13 Cold and Warm Fronts.

Describing the geographic pattern of tornadoes, which the Bible calls whirl winds, and violent low-pressure centers in a current temporal context is important because the Bible presents narratives in which they are used as metaphors for social and cultural disruptions caused by a wrathful deity. In other places, their depictions are clearly used to describe atmospheric conditions as part of environmental settings. Precipitation producing low-pressure centers occur across a wide expanse of space and are found on every continent; however, they are lower in frequencies in southwestern coastal regions where cold-water currents flow along their shorelines. In a global context, tornadic activity is rather rare, occurring mostly within a belt of latitudes that stretch from 30° N to 60° N and primarily in North America where highly contrastive air masses collide. Cold, dry air masses, especially in the spring and early summer months, push down from Canada and collide with warm moist air from the Gulf of Mexico. Fronts form along their points of contact, and in them, troughs and ridges produce low and high-pressure centers, respectively. Tornadic activity needs intense low pressure with an upper level wind shear to spur on instability in the atmosphere. Low-pressure centers form in troughs. Those conditions are currently more prevalent in North America; they are extremely uncommon elsewhere, although they sometimes occur in similar latitudes in the southern hemisphere.[24]

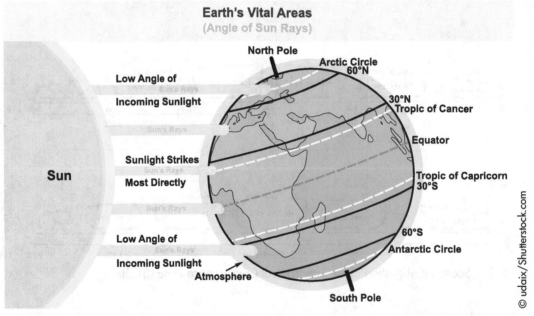

FIGURE 5.14 Most of the World's Landmass and People are between 30° and 60° N Latitude.

---

[24] Vann, *Forces of Nature*, 182.

ATMOSPHERIC FRONT

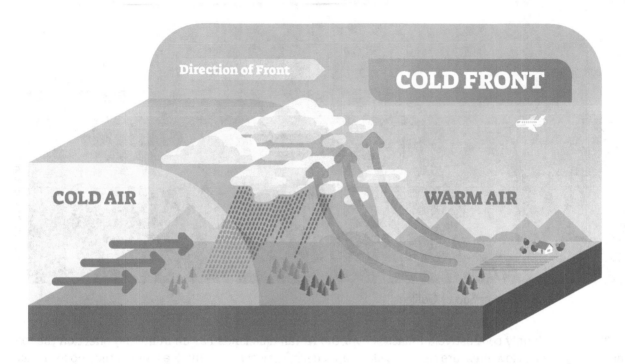

FIGURE 5.15 Cold and Warm Fronts.

FIGURE 5.16 Storm over Sea of Galilee.

The lack of highly contrastive air masses over the Abrahamic lands has an inhibiting effect on the formation of powerful low-pressure centers called super cells; they spawn tornadoes. When they do form, like the one that uprooted some trees in western Galilee in the spring of 2006, they are most often referred to as mini tornadoes by regional climatologists like Baruch Zvi of Tel Aviv University.[25] However, to date there is no clear definition of what conditions constitute a mini tornado. Even clear weather, heat-generated dust devils have been called mini tornadoes. True tornadoes, however, are products of larger circulations or low-pressure centers with upper level wind shears caused by aggressive cold air masses, and as such, they are potentially destructive to life and property. Based on photographic and damage evidence, the 2006 mini tornado in western Galilee was produced by a larger low-pressure center, so it was a rare, yet genuine Enhanced Fujita (EF)0 tornado. Aside from the "furious squall" on the Sea of Galilee that is described in Mark 4:35 to 41, imagery of tornadoes and the damage they caused are absent in the New Testament but sporadic in the Old Testament, which suggests temporally different weather patterns over the 1,000 plus years the Jewish texts were written. Scriptural references and contexts concerning atmospheric conditions are presented next.

---

[25] Ynetnews, "Mini Tornado Sweeps."

CHAPTER 6

Weather and Climate in Biblical Geotheology

INTRODUCTION

As the highest form of creation, biblical writers saw humanity as inextricably tied to a benefits-reward scenario initiated by God. All other biblical creatures are unable to elicit God's wrath or reward for moral or immoral thoughts and public behaviors, although the Genesis account of the fall of man involved a serpent who was likewise punished by God (3:14). Romans 8:18 to 22 speaks of the groaning of creation due to societal corruption. Most alterations from normal atmospheric conditions were depicted as the result of judgments of God on leaders and their wayward people, towns, and nations; however, Jesus's reaction to a furious thunder storm on the Sea of Galilee suggested it was part of the natural goings on and not enough to keep him awake.[1] Whereas the water cycle is used by God as a means to bring forth life where there was none, He also used the forces of violent storms to punish those who displeased Him. In Psalm 83:14 to 15 (New King James Version), the author calls upon God to punish his enemies with a storm, which indicate the presence of either convective storms or troughs formed along a cold front: "As the fire burns the woods, and the flame sets the mountains on fire. So pursue them with your tempest, and frighten them with your storm."

Now that we have established the manner in which geotheology can be used to appreciate how basic weather events were understood, let us briefly shift our attention toward atmospheric conditions. Whereas dew and frost can occur throughout the year in places with adequate humidity and would hence not be particularly noteworthy, tornadoes occur when there are strong frontal storms. Low-pressure centers that produce tornadoes require warm moist air in opposition to aggressive masses of cold air with upper level wind shears. Dew, on the other hand, just needs sufficient humidity, declining air temperatures, and cool objects on which to coalesce.

Frost and Dew. Frost is mentioned in several Old Testament books: Exodus 16:14; Job 38:29; Psalm 147:16; Jeremiah 36:30; and Zachariah 14:6. The presence of frost indicates that there was sufficient amount of moisture in the air to allow condensation to occur on cool grass blades, bushes, rocks, and the ground. Because cold air has less ability to hold gaseous water, it could be that frost was at least as common as dew. In Hosea 14:5, the author states "I will be like dew to Israel; he will blossom like a lily; like a cedar of Lebanon, he will send down his roots." In Micah 5:7, it is declared, "The remnant of Jacob will be in the midst of many people like dew from the LORD, like showers on the grass, which do not wait for anyone or depend on man." While biblical era residents collected dew from cool rocks during the drier months, there

---

[1] Matthew 8:23–27; Mark 4:35–41; Luke 8:22–25.

are many days in Jerusalem today in which there is insufficient humidity and cool temperatures in the summer months to allow water droplets to form. Rainfall itself is estimated to have been 50 percent more in the Dead Sea area, which was known as the Salt Sea, during the biblical era.[2] It is probable that dew often formed on the rocks of Masada, the ancient lofty fortress that overlooked the Salt Sea.

FIGURE 6.1 Drops of Dew on Grass.

FIGURE 6.2 Frost on Grass.

---

[2] Wolf, *Hydropolitics along the Jordan River*, 14.

FIGURE 6.3 Masada Looking toward the Salt Sea (Dead Sea).

It is worth revisiting the climate variations of the Salt Sea area that was briefly mentioned in Chapter 1; Scripture found in Ezekiel seems to validate the finding of a host of paleoclimatologists. Paleoclimatological evidence from the Near East suggests that arid conditions around the Dead Sea have alternated with wetter periods throughout the last 9,000 years.[3] In the later chapters of the Book of Ezekiel, the author describes hopeful prophesies for Judah and Jerusalem. In the 47th chapter, Ezekiel prophesies about a prolong period of time in which water flowing toward the Salt Sea grows from trickling stream into a river that he could not cross on foot. There is ample paleoclimatological evidence that the description he provides in 47:7–10 to 10 bears a striking resemblance to past geographies in the valley:

> When I returned, there, along the bank of the river, *were* very many trees on one side and the other.[8] Then he said to me: "This water flows toward the eastern region, goes down into the valley, and enters the sea. *When it* reaches the sea, *its* waters are healed.[9] And it shall be *that* every living thing that moves, wherever the rivers go, will live. There will be a very great multitude of fish, because these waters go there; for they will be healed, and everything will live wherever the river goes.[10] It shall be that fishermen will stand by it . . . there will be places for spreading their nets. Their fish will be of the same kinds as the fish of the Great Sea, exceedingly many."

The waters of the Jordan River create the Dead Sea as they flow into the basin. As can be seen in Figure 5.15, there are plains in the foreground. The River Jordan likewise has plains along its banks. In Genesis 13:10, Scripture describes how this land appeared around 2,000 BC when Abram and his brother Lot entered the valley. "And Lot lifted his eyes and saw all the plain of Jordan, that it *was* well watered everywhere (before the Lord destroyed Sodom and Gomorrah) like the garden of the Lord, like the land of Egypt as you go toward Zoar."

---

[3] Stein et al., "Abrupt Aridities and Salt Deposition," 567–575.

If precipitation was much more plentiful in Israel during parts of the biblical age, it is likely that it fell as both convective or heat generated events as well as frontal activity. One of the features of frontal activity is the creation of troughs and squall lines. Occasionally troughs form super cells and, if today's figures were true for the biblical age, one out of six supercells form tornadoes. Let us now shift our attention to biblical evidence of tornadoes.

Tornadoes. The word "whirlwind" was used by the writers of all of the Abrahamic texts to describe a weather event or as a metaphor for irresistible power, mass destruction, or lasting upheaval that modern writers associate with the destructive power of tornadoes. Still, it is not difficult to identify places in which the word is used in reference to a weather event that resembles a tornado. In 2 Kings 2:11 to 12, the prophet Elijah was taken up to heaven in a whirlwind. Because the event was witnessed by Elisha and Elijah's body was never found, it has been believed by members of faith communities that he never died. In Proverbs 37:8 to 10, the context of the use of the word whirlwind suggests that the writer had a clear understanding that these storms occur as a result of colliding air masses: "Out of the south cometh the whirlwind; and cold out of the north. By the breath of God frost is given: and the breadth of the waters is straitened." At least the portions of these books cited above were written between the fifteenth century BC and the sixth century BC. The clearest expressions of whirlwinds as weather events are found in Job and 2 Kings; they cover nearly 1,000 years; however, as is shown in the next section, there is reason to believe that climate during the time of Job was much colder and wetter than during the time of Jeremiah and certainly the writing of the New Testament.

FIGURE 6.4 Cold Front and Foreboding Squall Line.

FIGURE 6.5 Twister on a Lonely Road in the American Great Plains.

It should be pointed out that in Psalm 58:9; Proverbs 1:27; Jeremiah 30:23; Hosea 8:7 and Hosea 13:3; Amos 1:14; Nahum 1:3; Habakkuk 3:14; Isiah 17:13; and Zechariah 7:14 and 9:14 the word "whirlwind" is used as a metaphor for powerful, lasting upheaval, or destructive forces. Whirlwind as a metaphor suggests that the readers of those texts would have been familiar with the phenomenon as a destructive weather event.

While not as violent as whirlwinds, thunderstorms are nevertheless products of either frontal activity or heat-caused convection involving humidity. They occur today in the Saudi Peninsula, which would include the ancient Kingdom of Edom, but they are low in frequency and exceedingly uncommon outside of the south, especially the southwest away from where Edom was located.[4] In contrast to the current meteorological pattern, Job 28:26 claims that God made a law for rain and a path for the thunderstorm. In Isaiah 54:11, lashing storms tormented a city, and in Zechariah 7:14, destructive storms are described as coming out of the South. The New Testament records at least two instances in which thunderstorms are described by men traveling in boats: Mathew 8:23 to 27, Mark 4:35 to 41, and Luke 8:22 to 25 tell the story of Jesus calming a storm while sailing across the Sea of Galilee; Acts 27:13 to 44 tells of a frightening storm experience that the Apostle Paul had on board a ship sailing for Rome.

---

[4] Schwehdi, "Thunderstorm Frequency and Distribution," 252.

FIGURE 6.6 Jesus about to Calm the Storm on the Sea of Galilee.

Hail. Hail is a product of highly unstable low-pressure centers that form cumulonimbus clouds associated with convective and frontal storms. Hail is formed when condensation nuclei and coalescing water droplets become super cooled. As growing ice particles or graupel, they are thrusted upward and out of an intense low-pressure center. Like a tornado, hail producing storms are unusual in biblical lands today. During biblical times, especially in Job, Exodus, Joshua, and some of the Psalms, they are among the more commonly mentioned weather events in Scripture.

In Exodus 9:13 to 18, God tells Moses to deliver a warning to Pharaoh that if he does not release His people, God will send a devastating hail storm. Pharaoh refused, so Moses stretched out his rod toward heaven. "And the hail struck throughout the whole land of Egypt, all that was in the field, both man and beast; and the hail struck every herb of the field and broke every tree of the field" (Exodus 9:25). Later in Joshua, God uses a hailstorm to defeat the enemies of Israel. "And it happened, as they fled before Israel and were on the decent of Beth Horon, that the Lord cast down large hailstones from heaven on them as far as Azekah, and they died. There were more who died from the hailstones than the children of Israel killed with the sword" (10:11).

FIGURE 6.7 Hail.

FIGURE 6.8 Hail Damage on this Car's Windshield Resembles Bullet Holes.

In Job 38:22, the author begins a series of questions about the majesty of God by identifying various forces of nature, including storehouses of snow and hail, which God reserves for times of trouble, war, and battles. In the Book of Psalms, hail is mentioned as a meteorological event in a number of places (18:12; 78:47; 105:32; 148:8). In Psalm 78:47, the author declares that God destroyed their vines with hail and their sycamore figs with sleet. As in Exodus 9:25 and Joshua 10:11, hail is depicted as a geotheokolasis event.

Snow, Ice, and Frozen Conditions. As one can see from the average January temperature ranges in those locations, it is likely that the authors of the sacred texts, if they experienced similar conditions, would regard icy, frozen conditions as unusual situations and not particularly applicable to their audiences who would not have been able to appreciate them as metaphors; indeed, they would have been viewed as aberrant and resulting from acts of geotheokolasis. In surveying the Bible for evidence of cold, snowy, or icy weather, the Book of Job is rife with such imagery. Job even describes frozen bodies of deep water, which clearly indicate protracted cold spells; moreover, those conditions are not discussed in a way that suggests atypical weather or climate as in geotheokolasis context. In Job 37:10, the author wrote that "The breath of God produces ice, and the broad waters become frozen." In Job 38:30, a more general observation of cold weather is given: "When the waters become hard as stone, when the surface of the deep is frozen." This same passage taken from the more poetic King James Version states that "The waters are hid as with a stone, and the face of the deep is frozen." In Psalm 78:47, the author described how God used hail to destroy Egyptian vines and frost to kill sycamore trees. The author, who probably lived sometime around 1010 BC to 970 BC, was describing events that had occurred during the time of Moses. Hail and icy conditions associated with a formidable cold front is further described in Psalm 147:17: "He hurls down his hail like pebbles. Who can withstand his icy blasts?" Further evidence of a cold regional climate at the time of the penning of the Book of Job is found in 38:29, where the author rhetorically asks, "From whose womb comes the ice? Who gives birth to frost from the heavens?" All references in the Holy Scriptures to frozen environmental conditions are found in Job. The same is true

FIGURE 6.9 Frozen Lake and Snow.

for ice (Job 6:16, 37:10, 38:29). Although conducted outside of the Near East, which was the case in the paleoclimate studies cited earlier, Perry and Hsu show that a "little ice age" lasted from 1250 BC to 750 BC.[5] Although not necessarily frozen conditions, it seems that the author of the book of Jeremiah understood that snowfall in Lebanon was sufficient to recharge aquifers. The quality of the water must have been exceptional because Jeremiah asked, "Will a man leave the snow water of Lebanon, which comes from the rock of the field [likely bubbling out in springs]? Will the cold flowing waters be forsaken for strange waters" (Jeremiah 18:14)?

Isaiah wrote about snow in such a way that it suggests that the wintery precipitation was usual and part of the water cycle that provided life to plants and farmers: "For as the rain comes down, and the snow from heaven, and do not return there, but water the earth, and make it bring forth and bud, that it may give seed to the sower and bread to the eater" (55:10). The author of Proverbs was so clearly convinced that snow was typical of winter months that he used it in a metaphor to show the incompatibility of honor and foolishness: "As snow in the summer and rain in harvest, so honor is not fitting for a fool" (26:1).

As these passages illustrate, snow and icy conditions were observed in the time of Job but not in other Old Testament books. Icy conditions do not appear in biblical narratives after that time, but snow remains a fixture of passages written in the time of Solomon (ca. 971 BC–ca. 932 BC), Isaiah (ca. 700 BC), and Jeremiah (ca. 600 BC). From the book of Daniel (ca. 550 BC) onward through the time of Jesus and His disciples, most references to snow are used as an example of cleanliness or purity as in "white as snow" (Daniel 7:9). Because the use of snow as a metaphor for cleanliness does not tie directly to

---

[5] Perry and Hsu, "Geophysical, Archaeological, Historical Evidence," 12435.

© AlexZaitsev/Shutterstock.com

FIGURE 6.10 Snow has a Cooling Effect Due to Its High Solar Reflectivity or Albedo.

an environmental condition as depicted in the narrative, it is unclear if "white as snow" was just an expression passed on to succeeding generations or whether it was a commonly witnessed environmental condition. Interestingly enough, frost disappears from the Scriptures after Jeremiah. There are no accounts of snow (as a precipitation event), icy conditions, or even frost in the New Testament. Judging from the sacred passages, it seems that climate was much cooler and wetter in the time of Job, and then a gradual warming trend followed, perhaps intermittently; so, by the time of Darius, snow and frost imagery are lost from the texts. This certainly does not mean that frost or snow never fell again in the Near East; it simply suggests that their impact must have been light because little or no attention was given them. Nonetheless, the presence of icy conditions that produced frozen rivers as a routine occurrence were more likely to occur in the Middle Bronze Age Cold Epoch (1800 BC–1500 BC) than they were during the warmer periods that followed, including a return to a colder climate during the Iron Age Cold Epoch. This notion is substantiated by references to weather and climate conditions in other Old Testament texts. While those books contain narratives of a colder climate than is found in the Levant today, those conditions are much milder than those depicted in Job. The Book of Job is arguably one to four centuries older than most current estimates place it, although a few writers place Job's authorship in the time of the Patriarchs (ca. 2200 BC). From a paleoclimatological perspective, that date does not work. For instance, the only aberrant climate epoch from 3000 BC to 2000 BC was the extensive drought period known as 4.2 Kiloyear BP Aridification Event that occurred during the twenty-second century BC.[6]

---

[6] Bar-Mathews and Ayalon, "Mid-Holocene Climate Variations," 168.

Assessing the weather patterns. In Tables 1 and 2, weather events associated with atmospheric heating and cooling are shown with respect to their locations among the 66 books in the Bible. Passages in which whirlwinds are used as metaphors for some sort of upheaval are not included. Weather events such as dew and thunderstorms are found in the greatest number of books (14 and 10, respectively). At the other extreme are ice and frozen conditions; they are found only in the Book of Job, which is the only book in which each of the various weather events depicted in the tables can be found. The only New Testament books in Table1 are Luke, Mathew, and Acts; Luke and Mathew recount the same story of Jesus calming a tempest on the Sea of Galilee. The story in Acts, too, is set at sea, so the perception of its power was most likely affected by being tossed about in a ship designed by ancient mariners.

TABLE 1 Nonmetaphoric Weather in Biblical Texts

Weather Condition	Biblical Text New International Version (NIV & NKJV)
Dew	Genesis 27:28; 27:39 Exodus 16:13–14 Numbers 11:9 Deuteronomy 33:13; 33:28 Judges 6:37–40 2 Samuel 1:21 1 King 17:1 Job 29:19; 38:28 Psalm 133:3 Proverbs 3:20 Songs of Songs 5:2 Isaiah 26:19 Daniel 4:15 ($N = 13$) Haggai 1:10 Zechariah 8:11–13 ($N = 14$)
Thunder, Lightning, Storm,	Exodus 9:23–24, 28–29, 33–34; 19:16; 20:18 1 Samuel 7:10; 12:17–18 2 Samuel 22:15 Job 9:17; 28:26; 30:22; 36:29–30, 32–33; 37:3, 11, 15; 38:24–25, 35 Psalm 18:12, 14; 55:8; 83:15; 77:17–18; 78:48; 97:4; 105:32; 107:29; 135:7; 144:6; 148:8 (Not analyzed) Proverbs 10:25 Isaiah 4:6; 25:4; 54:11 Jeremiah 10:13; 11:16; 23:19; 25:32; 30:23; 51:16 Ezekiel 1:4, 30:16 Amos 1:14 Jonah 1:4 Luke 8:23–25 (Not analyzed) Mathew 8:24 (Not analyzed) Acts 27:18 (Not analyzed) ($N = 10$)
Floods	Genesis 6:17, 7:6–7, 10–11, 17, 24; 8:2; 9:11, 15, 28 Psalm 69:2, 15 Joshua 3:15; 4:18 Job 20:28; 22:11, 16 ($N = 3$)
Hail	Exodus 9:13;18–19; 22–26; 28–29; 33–34; 10:5,12,15 Joshua 10:11 Job 38:22 Psalm 18:12; 78:47–48; 105:32; 147:17; 148:8 (Not analyzed) Isaiah 28:2 ($N = 4$)

Weather Condition	Biblical Text New International Version (NIV & NKJV)
Tornadoes (whirlwinds)	2 Kings 2:1 (Elijah), 11 Job 37:9 Psalm 77:18 (not analyzed) Isaiah 21:1; 40:24; 41:16 Jeremiah 23:19; 25:32; 30:23 Ezekiel 1:4 Amos 1:4 Hosea 4:19 Nahum 1:3 Zechariah 7:14 (N = 9)
Frost	Genesis 31:40 Job 38:29 Psalm 78:47; 147:16 Jeremiah 36:30 (N = 3)
Snow	2 Samuel 23:20 1 Chronicles 11:22 (same as 2 Samuel 23:20); (only one book is included in the analyses) Job 6:16; 24:19; 37:6; 38:2 Psalm 147:16; 148:8 Proverbs 31:21 Isaiah 55:10 Jeremiah 18:14 (N = 5)
Ice and Frozen Conditions	Job 6:16; 37:10; 38:29–30 (n = 3) Psalm 147:17 (refers to objects that were most likely hail)

Statistical Patterns. The chronology of geotheological narratives is presented in Table 2. The many books found in Psalms were penned over an estimated 960-year-period of time (1410 BC to 450 BC). Because precise dates for many of the writings in the Book of Psalms cannot be stated with a degree of certainty, the Book of Psalms is not listed in Table 2. Passages in which whirlwinds and other weather events such as storms and hail are used as metaphors for some sort of social or political upheaval were not included. Because of an interest in analyzing real weather events, passages in which weather-like phenomena are mentioned in a geoteleological or prophetic (future) context were excluded from the statistical analysis. Because of its unique weather patterns, the book of Job was not included in the statistical analyses.

The ordering of books was based on their assumed age. This ordering facilitated a comparison of three distinct time periods in biblical history: 1440 to 970 BC; 950 to 560 BC; and 550 to 423 BC.

To look for temporal patterns of volatile and cold weather events, chi square tests were conducted. The tests were used to compare the frequencies of weather events resulting from unstable air masses and cold conditions (storms, hail, tornadoes, and snow). Comparisons of weather frequencies were made among books thought to have been authored between 1440 and 970 BC ($n = 9$), those written between 950 and 560 BC ($n = 19$), and the texts written between 550 and 423 BC ($n = 9$). First, a chi square was performed to determine if there was a significant difference among the three distinct time periods, and a significant difference was found, ($X^2[1] = 12.14$, $p < .01$). (See Table 2.) Having established that there was a significant difference among the three time periods, pairwise comparisons were then made. A comparison of books written between 1440–970 BC and 950–560 BC resulted in a significant chi square, ($X^2[1] = 4.84$, $p < .05$). Likewise, a comparison of books written between 950–560 BC and 550-423 BC resulted in a significant chi square, ($X^2[1] = 8.88$, $p < .01$). The last pairwise comparison was made between the years 1440–970 BC and 550–423 BC. The statistically insignificant result of that test suggests that the relatively warm climate of the 1440–970 BC period returned during the 550–423 BC epoch, ($X^2[1] = 1.06$, $p > .05$).[7]

TABLE 2 Old Testament Books by Dates and Weather Events by Type ($N = 37$)

Book	Est. Date[8]	Storms	Hail	Tornadoes	Snow
Job	1500 BC–200BC	1	1	1	1
Genesis	1440 BC	0	0	0	0
Exodus	1400 BC	1	1	0	0
Leviticus	1440 BC	0	0	0	0
Numbers	1400 BC	0	0	0	0
Deuteronomy	1400 BC	0	0	0	0
Joshua	1375 BC	0	1	0	0
Judges	1050 BC	0	0	0	0
Ruth	1050 BC–500 BC	0	0	0	0
Song of Songs	970 BC–930 BC	0	0	0	0
Proverbs	950 BC–720 BC	1	0	0	1
*Ecclesiastes	931 BC	0	0	0	0
1 Samuel	931 BC–722 BC	1	0	0	0
2 Samuel	931 BC–722 BC	1	0	0	1
Obadiah	586 BC	0	0	0	0
*Joel	835 BC–805 BC	0	0	0	0
Jonah	760 BC–612 BC	1	0	0	0
Amos	760 BC–750 BC	1	0	1	0
Hosea	750 BC	0	0	1	0
Micah	704 BC–696 BC	0	0	0	0
Isaiah	700 BC–690 BC	1	1	1	1
Nahum	612 BC	0	0	1	0
Zephaniah	630 BC	0	0	0	0
Habakkuk	600 BC	0	0	0	0
Ezekiel	593–587 BC	1	0	1	0
Lamentations	587 BC	0	0	0	0
Jeremiah	626 BC–586 BC	1	0	1	1
1 Kings	560 BC–538 BC	0	0	0	0
2 Kings	560 BC–538 BC	0	0	1	0
Daniel	550 BC	0	0	0	0
Haggai	520 BC	0	0	0	0

[7] In these tests, frequency data were tabulated based on the following criteria: weather events were totaled for a time period. For example, in the 1440–970 BC category there were 36 cells (9 books times 4 categories). There were 3 weather events and 33 non weather events. In category 950-560 BC, there were 76 cells (19 x 4). A total of 20 weather events were identified, leaving 56 empty cells. In category 550–423 BC, 1 weather event was identified, so there were 35 empty cells.

[8] Dates for the Tables were taken from Lee, *The American Patriots Bible*. *There are a number of scholars who place the dates for Ecclesiastes and Joel in the postexile era. See Fox, *The JPS Bible Commentary*, x; Allen, *The Books of Joel, Obadiah, Noah & Micah*, 231–232.

Book	Est. Date[8]	Storms	Hail	Tornadoes	Snow
Zechariah	520 BC–475 BC	0	0	1	0
Ezra	538 BC–457 BC	0	0	0	0
1 Chronicles	425 BC–400 BC	0	0	1	0
2 Chronicles	425 BC–400 BC	0	0	0	0
Esther	465 BC	0	0	0	0
Malachi	450 BC	0	0	0	0
Nehemiah	423 BC	0	0	0	0
Total		10	4	9	5

The book of Job was not compared in the chi-square tests because frozen conditions are uniquely mentioned in it. Ice is named in Psalm 147:17, but it is in reference to casting forth hail as "ice-like morsels." In Job, on the other hand, it is clear that the author is describing frozen bodies of water (38:30). Even biblical texts written during the Iron Age Cold Epoch, which lasted from ca. 900 BC to ca. 300 BC, do not feature descriptions of frozen bodies of water.[9] However, cold front-produced weather events like storms, tornadoes, snow, and hail are frequently mentioned ($n = 20$) in those books. If the Pentateuch and other Davidic books were written between 1500 BC and 970 BC, which is the assertion made by more conservative scholars like Richard D. Lee, Job shares little in common with them relative to very cold weather events. Aside from the Genesis flood, the most extreme cold weather mentioned in the Pentateuch is frost, appearing in Genesis 31:40. Storms and hail, however, are mentioned in other premonarchic books. While snow also appears in Job, it does not appear in the Pentateuch. The range of weather events depicted in Job, therefore, suggest an earlier period and may well have been written around 1800 BC when a period of "strong cooling" occurred.[10] Based on the lack of cold weather narratives in the Pentateuch and the other books written before 970 BC, it is arguable that they were most likely written between the Middle Bronze Age Cold Epoch (1800 BC–1500 BC) and the Iron Age Cold Epoch (900 BC–300 BC). Given that a cold period existed from 1300 BC to 1124 BC,[11] the Pentateuch may well have been written in the 1400s as suggested by those who argue for an older authorship.

The weather events depicted in the books written between 1440 BC and 970 BC are similar to those books thought to have been written between 550 BC and 423 BC. Given that the Iron Age Cold Epoch included the years 550 BC to 423 BC, with an exceptionally cold period running from 500 BC to 450 BC,[12] it is likely that those books were written closer to the period known as the Roman Warm Period (300 BC–550 AD).[13] The passage in Zechariah 7:14 describes a situation in which a tornado wreaked havoc on people and lands; the people were scattered among the nations while a pleasant land was left desolate. Zechariah was most likely written between 520 BC and 475 BC when aggressive cold fronts created atmospheric instability.[14]

SUMMARY

The ethnomethodological delineation of geotheological imagery presented in this paper is not an exhaustive description of the ways in which Abrahamic leaders viewed the interplay between nature and the worship of God. They collectively provide an overview of the basic ways in which some ancient peoples understood

---

[9] Gribben and Lamb, "Climate Change in Historical Times," 68–82.

[10] Ward, "Climate of the Holocene."; Siklosy et al., "Middle Bronze Age Climate Change."

[11] Drake. "The Influence of Climate Change," 1862.

[12] Lamb, *The Changing Climate*, 6, 48, 167.

[13] Wang et al., "Seasonal Climate Change Across the Roman," 230.

[14] Lamb, *The Changing Climate*.

weather and climate. Cold weather patterns appear in the oldest texts. They are absent in the more recent writings; hence, climate change is evident in Scripture. This is important because much has been made about anthropogenic causes of climate change through recent industrialization.

The ethnomethodological and geotheological approaches used to frame climate-related imagery scattered throughout the Bible show that from ca. 1800 BC to 300 AD the Near East was a wetter, warmer, and at times, cooler and atmospherically unstable place to live. The past geographies expressed in scriptural narratives show that the Book of Job should be dated to the Middle Bronze Age Cold Epoch (1800 BC–1500 BC). The Pentateuch shows weather and climate conditions that would be consistent with a more moderate climate that was sandwiched between the Middle Bronze Age Cold Epoch and the Iron Age Cold Epoch (900 BC–300 BC). The post-Babylonian exile books, which feature two tornadoes, are consistent with the overall pattern of weather that is found during a transition from the Iron Age Cold Epoch to Roman Warm Period. Given that tornadoes and other cold weather events are not common in the region today, it is clear that climate is warmer and drier now than it was before the Roman Warm Period. These data show that the Bible is a relevant and useful document in discussions on climate change.

CHAPTER 7

Quranic Images of Weather and Climate

INTRODUCTION

Much of the Saudi peninsula is dry and sandy. Humidity reaches its highest levels near the Red Sea and the Gulf of Aden.[1] Saudi Arabia, the largest of the six countries that occupy the peninsula, is characterized by a hot desert climate.[2] Major cities like Mecca, Medina, Riyadh, Jeddah, and Hofuf, which were functioning settlements during the time of Muhammad, are located in desert lands with relative humidity that can drop to as low as 2 percent to 3 percent.[3] While there is enough precipitation in the southwest to support sparse grassy vegetation, the amount of moisture lost through evapotranspiration is greater than the amount received as precipitation during the year. These contemporary weather and climate descriptions take on greater importance when we take a look at the kinds of atmospheric conditions that are depicted in the Quran.

Muhammad lived in the city of Mecca as a boy and young man. As an adult, he moved northeastward to a city called Yathrib, which was renamed Medina (The City of the Prophet). Most of Muhammad's life was spent on the Saudi Peninsula. Muslims believe that the Quran, which means "recitation," was revealed to Muhammad (571–632) over a 22-year period of time. His spiritual journey and the composition of the Quran began while he was seeking a cool respite in the Cave of Hira (see Figure 7.2). Muslims believe that it was in the cool of the cave that the Archangel Jibril (Gabriel) appeared to Muhammad and instructed him to write down and recite what he was to be shown and told. Many of the Quran's stories and themes are loosely based on events and people depicted in the Bible (the Taurat for Torah and the Injil for the gospel of Isa or Jesus). However, a significant amount of material in the Quran is expressed as new, prophetic information; nevertheless, for readers conversant with the Bible, there are familiar themes in the Quran: evil, punishment, reward, and environmental forces that bend to the will of God. However, there are serious and profound theological differences

---

[1] Patlakas et al., "Regional Climatic Features," 220.

[2] Bahrain is an independent, island country that many observers also count among seven Saudi Peninsula states. Politically and culturally it makes sense to include Bahrain among the seven; however, for our purposes, we are looking at weather and climate on the Saudi Peninsula. A hot desert climate is designated BWh on the Köeppen-Geiger climate classification system.

[3] Climate-Data.org, "Climate Saudi Arabia."

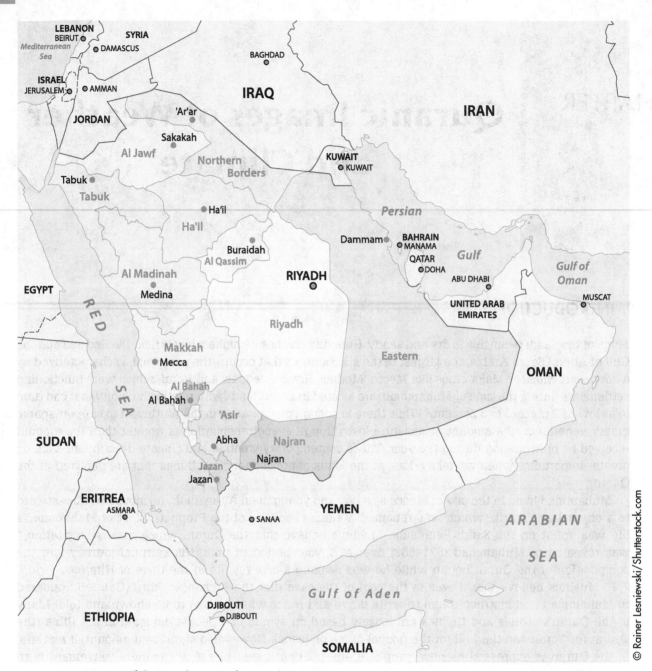

FIGURE 7.1 Map of the Saudi Peninsula.

among the three Abrahamic faiths (Judaism, Christianity, and Islam). Exploring those differences is beyond the scope of this book, but if you are interested in a deeper theological emersion in Islam and the Quran, I invite you to read my *Puritan Islam: The Geoexpansion of the Muslim World*. Our focus in *Climate Change in History* is on the ways in which the Quran discusses and describes weather and climate events. This is a fruitful exercise because it helps us to gauge the kinds of climate changes that have taken place in this part of the Levant and the rest of the Saudi Peninsula during the seventh century AD. Because a writer's geographic imagination can affect geotheological imagery in any text, narratives in the Bible and Quran can and should be seen in light of paleoclimatological studies.

© Rainer Lesniewski/Shutterstock.com

FIGURE 7.2 Cave of Hira.

In a white paper summarizing the most important and relevant studies on Levant and Saudi Peninsula climatology during the time of Muhammad and the writing of the Quran, Michael McCormick and a host of his colleagues describe the regional climate. As McCormick et al. write, "North Atlantic sea ice started to advance at that time, and the region began to cool again. A cooling trend persisted until c. 800, although no clear evidence of a corresponding increase in regional precipitation exists at present."[4] Nevertheless, McCormick et al. go on to point out that regional flood events occurred in 667, 716, 724/25, and 740.[5] The white paper also lists 11 regionally harsh winters: "609/10, 623, 628, 662 (late frost), 667/68, 668–670, 683/84, 698/99, 707, 715, 717/18."[6] Logic suggests that more precipitation would have been produced due to the cooling trend. As a consequence of colliding air masses, atmospheric uplift can create volatile storm events and perhaps flooding. Unlike the Bible, which was written over a wide expanse of time, the Quran was written from 609 AD to Muhammad's death in 632 AD; the Quran, it would seem, was written during a cooling trend that featured harsh winters that no doubt created atmospheric instability as warm air masses were pushed by aggressive cold fronts. Protracted warm frontal events most likely produced a good bit of surface water flow and the formation of ponds and lakes. Keep in mind that there are no permanent water bodies like rivers, ponds, and lakes in Saudi Arabia. There are dry river beds called wadis that can flood during convective storm events. As is shown in the pages that follow, geotheological imagery in the Quran certainly support the occurrence of the kinds of unstable weather events that would be produced by a cooling trend like the one described in McCormick's white paper.

---

[4] McCormick et al., "Climate Change during and after the Roman Empire," 202.

[5] Ibid.

[6] Ibid.

FIGURE 7.3 Enhanced Satellite Image of the Saudi Peninsula.

FIGURE 7.4 Wadi Aqaba Tabuk, Saudi Arabia.

Quranic Geotheology

In Surah 2, Muhammad recounts the creation of Adam. In his original state, Adam was healthy and obedient to Allah. His well-watered, lush, pristine garden provided for all of his physiological needs, but in verses 35 to 36, Allah is seen giving Adam directions on what and indeed what not to eat. These directions were apparently not followed, for the Quran swiftly declares that Adam and his unnamed kin were immediately banished from the abundance of Allah's bountiful garden. The Quran's retelling of the Genesis account of creation and Adam's original sin is rather brief. Humanity's flawed condition, which results from selfish choices and actions, has plagued humankind ever since, and it is in the creation story that the Quran establishes beliefs in geotheokolasis (God, nature, punish) and geotheomisthosis (God, nature, reward) as the basis of humanity's relationship to God. In Islam, to say that the Quran uses fire and water as the basis for kolasis (punishment) and misthosis (reward) is not an understatement. Perhaps nowhere in the Muslim holy book is this more apparent than in Surah IV: "Allah has promised to the believing men and believing women gardens, beneath which rivers flow, to abide in them, and goodly dwellings in gardens of perpetual abode; and best of all is Allah's goodly pleasure—that is the grand achievement (9:72)" Not content with resting on a sacred promise, verse 9:73 qualifies the lush image of an earthly natural paradise for those deserving of Allah's good will and grace by issuing this command: "O Prophet! Strive hard against the unbelievers and the hypocrites and be unyielding to them; and their abode is hell, and evil is the destination." There are many places in the Quran in which the unbelievers' evil destination is described. Suffice it to say that unbelievers are punished on both sides of the grave, and the forces of nature often play major roles in dispensing Allah's decrees on this side of eternity.

As in the Bible, God is seen as instructive and willing to use environmental forces to punish unbelieving men and women, or, conversely, provide believers with gardens and life-giving waters. For many years in the United States, some Christians have believed that storms and other destructive events in nature were regarded as "acts of God." The phrase has become part of the legal lexicon of American jurisprudence.[7] It may be one of the few existing concepts associated with secular life in America that the fundamentalist Christian and Islamic communities will recognize and support.

© Vuk Kostic/Shutterstock.com

FIGURE 7.5 Each of the Abrahamic Religions Acknowledge the Garden of Eden.

---

[7] Henry Campbell Black, *Black's Law Dictionary*, 33.

Nevertheless, to the faithful Muslim and his *umma* (community) both rewards and punishments bequeathed by nature are authorized by Allah. "And with Him are the keys of the unseen treasures—none knows them but He; and He knows what is in the land and the sea; and there falls not a leaf but He knows it, nor a grain in the darkness of the earth, nor anything green nor dry but [it is all] in a clear book."[8] Muhammad went on to caution the faithful: "Do you not see that Allah created the heavens and the earth with truth? If He please He will take you off and bring a new creation, and this is not difficult for Allah."[9] Muhammad even warned himself that his place in Allah's world was not assured: "Maybe my Lord will give me what is better than your garden, and send on it a thunderbolt from heaven so that it shall become even ground without plant. Or its waters should sink down into the ground so that you are unable to find it."[10] This imagery suggests a capricious existence with little assurance of a mild climate. Geotheomisthosis and geotheokolasis narratives in the Quran conjure up images of lush gardens, rivers, fire, thunder, lightning, and even a tornado. Each is used to contrast the earthly fates of both believers and unbelievers. "For those who do good is good [reward] and more [than this]; and blackness or ignominy shall not cover their faces; these are the dwellers of the garden; in it they shall abide. And [as for] those who have earned evil, the punishment of an evil is the like of it, and abasement shall come upon them—they shall have none to protect them from Allah—as if their faces had been covered with slices of dense darkness of night; these are the inmates of the fire; in it they shall abide."[11]

As is the case with the Book of Genesis, the Quran features a retelling of the Noachian flood. According to the Quran, Noah warned others of Allah's wrath if they did not change their ways and submit (the literal meaning of Islam) to the teachings of Allah. "But they rejected him, so We delivered him and those with him in the ark, and we made them rulers and drowned those who rejected Our communications; see then

FIGURE 7.6 Reconstruction of Noah's Ark.

© photostockam/Shutterstock.com

---

[8] Surah 6:59.

[9] Surah 14:19–20.

[10] Surah 18: 40–41.

[11] Surah 10:26–27

what was the end of the [people] warned."[12] Unlike the Genesis account, the Quran teaches that nonfamily members survived the flood event and became Noah's subjects.

Images of Rain, Storms, Tornadoes, and Cold Fronts

The Quran features a number of geotheological narratives that show life-giving water coming down from clouds and flowing through rivers while nourishing green gardens. As noted earlier, there are no permanent flowing or stationary water bodies in Saudi Arabia. Notice how Muhammad describes this aspect of the hydrologic cycle: "Who made the earth a resting place for you and the heaven a canopy and [Who] sends down rain from the cloud, then brings forth with it subsistence for you of the fruits; therefore, do not set up rivals to Allah while you know."[13] The kind of rainfall that is described in 2:22 is typical of the slow passage of a stationary front; rainfall during a stationary front is relatively gentle and steady and often lasts for more than a day. The Quran presents this kind of a meteorological event as a reward for the dutiful believer: "But as to those who are careful of [their duty to] their Lord, they shall have gardens beneath which rivers flow, abiding in them; an entertainment from their Lord, and that which is with Allah is best for the righteous."[14]

Muhammad was convinced that water from the sky was a blessing from Allah. In all of the bounty found in Allah's world, Muhammad insisted that the creator's divine hand can be seen. Water, the most precious commodity on earth, gives life to the earth after its death. The Quran promises that there is a sign in the hydrologic cycle for those who listen.[15]

FIGURE 7.7 Hydrologic Cycle.

©Artisticco/Shutterstock.com

---

[12] Surah 10:73. In many parts of the Quran, Muhammad uses "We" which suggests that he and God acted together in dispensing punishments.

[13] Surah 2:22.

[14] Surah 3:198.

[15] Surah 10:65.

Whereas the water cycle is used by Allah as a means to bring forth life where there was none, He also uses the forces of violent storms to punish those who displease Him. "When Allah intends evil[16] to a people, there is no averting it, and besides Him they have no protector. He it is Who shows you the lightening causing fear and hope and [Who] brings up the heavy cloud. And the thunder declares His glory with His praise, and the angels too for awe of Him; and He sends the thunderbolts and smites with them whom He pleases, yet they dispute concerning Allah, and He is mighty in prowess."[17] In a message that is

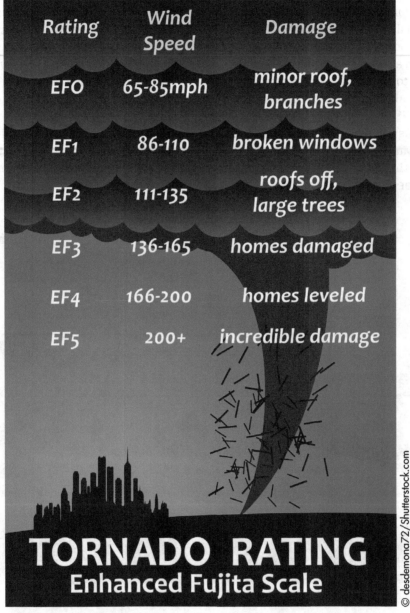

FIGURE 7.8 Enhanced Fujita Scale (The Enhanced Version Features Types of Damage).

---

[16] The use of the word "evil" to describe an act of God is unique to Islam.

[17] Surah 13:11–13.

reminiscent of the refrain echoed in Job 1:21: "the Lord giveth and the Lord taketh away," the *Quran* tells of a tornado that destroyed a man's home, garden, and belongings. In this hypothetical narrative, we can see that there was flowing water, lush vegetation, and, most importantly, a distinctly different weather event that would be virtually impossible in today's Saudi Arabia. Although the narrative is hypothetical, the environmental setting was a scene with which Muhammad's followers would have been familiar: "Does one of you like that he should have a garden of palms and vines with streams flowing beneath it; he has in it all kinds of fruits; and old age has overtaken him and he has weak offspring, when, [lo!] a whirlwind with fire in it smites it so it becomes blasted; thus Allah makes the communication clear to you that you may reflect."[18] As in the Bible, a Quranic whirlwind is a tornado. Given the destruction seen in the narrative in 2:226, this event was not a mini tornado or dust devil. While it is possible that the use of "whirlwind" was hyperbolic, the nature of the destruction suggests a literal weather event; moreover, the writer would have had to possess real-world knowledge of whirlwinds to be so precise in describing the consequences of a tornado. Based on the description of the event in 2:226, the damage was the consequence of the passage of a supercell producing at least an Enhanced Fujita (EF)2 vortex (see the Enhanced Fujita Scale in Figure 7.3). Further on in Surah 17:68, Muhammad warns his followers that Allah could send a tornado on them for their destruction, so they should not feel secure. In theological terms, this is an example of a theology of deterrence. Keep in mind that "Islam" means "submission"; using fear to cause submission is not a problem in the Islamic world.

In Surah 6:6, Muhammad, who once again uses the pronoun We, as he expresses bewilderment that his generation does not think about how capricious life is in the hands of a judgmental Allah. In this passage, he notes that "We have destroyed before them, whom We had established in the earth as We have not established you, and We sent the clouds pouring rain on them in abundance, and We made the rivers to flow beneath them, then We destroyed them on account of their faults and raised up after them another generation." It is not clear that the cessation of rain was the instrument of punishment in 6:6, but rain was clearly a sign of Allah's benevolence and grace. With the limited technology of seventh century life, an agrarian and pastoral people, as most of the residents of the Saudi hinterland surely were, having sufficient rain to create surface water flow meant healthier vegetation and greater abundance.

It is important to recall that precipitation is caused by two kinds of atmospheric conditions. There is the heat generated or convective variety that occurs in the warmer months (see Figure 7.9). In these events, the sun warms the earth's surface and water bodies. Heat is reradiated back into the troposphere causing air to rise. If there is sufficient lift or instability, a rising air mass cools and, if there is enough humidity, condensation can occur. In this kind of rain event, precipitation is local, and the duration of the rainfall is often less than an hour. As we saw earlier, a warm front is a frontal precipitation event. Warm fronts produce protracted precipitation that may last for days. These gentle rains are highly desirable among farmers. The precipitation is not violent, does not disturb or erode loose soil, and percolates down into the root zones of soil. A cold front, on the other hand, typically produces rapidly rising and sometimes rolling masses of air. Precipitation is often intense. Hail may form and high winds with a squall line are produced. Tornadoes can also form. When the front passes, cold air settles in and there is often a significant drop in air temperatures. If the Saudi Peninsula in the seventh and eighth centuries experienced a cooling trend as suggested by McCormick et al., the Quran would provide examples of the passage of cold fronts. Such is the case in Sura 3:117: "The likes of what they spend in the life of this world is as the likeness of wind in which is intense cold (that) smites the seed produce of a people who have done injustice to their souls and destroys it" (see Figure 7.4). However, if there are frequent instances in which warm fronts bring steady rain, crops, and fruit bearing plants produce lush greenery and vegetation. In Figure 7.5, the image on the left is a warm front. The cold front on the right shows atmospheric instability.

---

[18] Surah 2:226.

FIGURE 7.9 Summer Afternoon Convection Causing the Formation of Cumulonimbus Clouds

FIGURE 7.10 Fast Moving Cold Front and Squall Line.

© stihii/Shutterstock.com

FIGURE 7.11 Contrasting Cold and Warm Fronts.

© Felix Friebe/Shutterstock.com

FIGURE 7.12 Wadi Disah in Tabuk Region, Saudi Arabia.

Flowing Water and Greenery

An obvious benefit of rain is flowing water that supports lush vegetation. Although Saudi Arabia has a number of wadis, it is important to keep in mind that they only contain intermittent streams and infrequent deluges of water. Most of the time, they are dry river beds. When there has been sufficient precipitation, the water table rises and is within reach of the tap root of trees, so tree stands that are too small to be called forests can become established. The same phenomenon occurs in the western Great Plains of the United States where river bottoms support stands of cottonwoods. The kinds of images that we will see in this section go beyond the vulnerable nature of the ecosystem in today's wadis. The fact that Muhammad and his followers recognized these kinds of lush settings tells us that flowing water and greenery were not unusual. Note that in Figure 7.5, the Wadi Disah's palmetto trees are green, but the shorter-rooted grass clumps are not well established; extensive patches of exposed soil are clearly visible.

In Surah 16:10 and 11, a setting is described that conjures up images of wine country in Italy or southern California. In those two verses, various plants species are mentioned: "He it is Who sends down the water from the cloud for you; it gives drink, and by it (grow) the trees upon which you pasture. He causes to grow for you thereby the herbage, and the olives, and the palm trees, and the grapes, and of all the fruits." In Surah 20:53 and 54, another plentiful, lush description is given. In this and other cases, Muhammad uses the pronoun "we" as he tells his followers that we "sent down water from the cloud; then thereby We have brought forth many species of various herbs. Eat and pasture your cattle." Later in the Surah 80:27 to 32, Muhammad describes a variety of plant species, including thick gardens, palms, grapes, various fruits, grains, and clover. The mentioning of clover is especially significant because "The best time to plant clover in the spring is from mid-April to mid-May to coincide with spring rains. Clover needs about 30 inches of rain, ideally at least 45 inches . . ."[19] Grape vines, too, need at least 30 inches of rainfall per year, unless they are irrigated. While too much summer rain can cause grey rot and other diseases, grape vines nevertheless need some summer rain to maintain vigor. Spring rains are most important, however. Incidently, there is no evidence that irrigation was a well-established practice in Saudi Arabia in the seventh century.

© Deyan Georgiev/Shutterstock.com

FIGURE 7.13 Grape Vines.

© Zeljko Radojko/Shutterstock.com

FIGURE 7.14 A Clover Field.

---

[19] Hoyt, "When to Plant Clover Seeds."

SUMMARY

As you may recall, the Saudi peninsula features a Köppen climate classification of BWh, which identifies a hot desert where evaporation rates outpace rainfall. It is also worth further noting that to be classified as a desert, whether hot, cold or mild, a location has to have less than 10 inches (250 cm) of rain a year. In cold latitudinal and continental locations and depending on the texture of snow flakes and sleet, 120 to 130 inches (3.04–3.30 meters) or less of snow per year marks a cold place as a desert. The geotheological images or narratives in the Quran provide quite a contrast to the kind of climate that prevails across the sandy peninsula. While there are no permanent flowing water bodies in Saudi Arabia today, the Quran is replete with imagery that mentions flowing rivers. Because wadis are river beds with intermitent flowing streams today, the kinds of vegetation mentioned in Quranic passages suggest that the rivers mentioned in texts are not of the wadis variety. Indeed, grapes and clover which are mentioned in the Quran as part of the local setting, require at least 30 inches of rainfall per year. The Quran, therefore, offers narratives that support the findings of the white paper written by McCormick et al.

© Vladimir Endovitskiy/Shutterstock.com

FIGURE 7.15 Arctic Tundra, Despite Visible Surface Water, is a Cold Desert.

CHAPTER 8

Economics, Politics, and Future Climates

INTRODUCTION

There is a common refrain among intellectuals and philosophers which claims that "Those who cannot remember the past are condemned to repeat it."[1] Normally that observation applies to human actions that affect cultural, social, and political movements. For instance, what lessons can we learn from the American Civil War that can help future generations to avoid such a calamity? When it comes to the environment, there are also many historical lessons that can teach us ways to avoid catastrophes. Among the disasters that come to mind are the Donora, Pennsylvania air pollution tragedy, the death and fragile rebirth of Lake Erie, and the American Dust Bowl.[2] While those disasters certainly involve atmospheric conditions, they do not necessarily point to long-term changes in weather. That does not mean that we cannot learn from past climate epochs. However, one of the lessons that we must learn is that we have to think objectively about climate change; it is extremely difficult to make a connection between humans and the Medieval Optimum (ca. 700–ca. 1100) and the Little Ice Age (ca. 1250–ca. 1850). The same is true for the cooling trend that lasted from around 600 AD to 800 AD. A volcanic eruption is thought to have contributed to an especially cold year in 1816 that historians call the year without a summer. Looking back further in time, there is the Roman Warm Period that lasted from around 300 BC about 400 AD. Theories exist to explain that warm period, but there is little evidence among those speculations that point to human activities. Despite these clear historical periods that suggest that nature itself was the agent that temporarily changed the climate, many modern observers maintain a view that was common among the ancients; we humans are responsible for all that befalls us. Scientists too sometimes invoke religious symbols in their descriptions of the perceived weather conditions. Geographer Wallace Akin claims that our gluttony (one of the seven deadly sins) for oil has caused our current climatic conditions and that we can expect more frequent and extra powerful tornadoes to occur in the future.

> I, as do most scientists, believe that global warming is increasing the frequency of severe storms, and this is a situation little understood by the public and their elected officials, who tend to focus only on temperature changes. But it is more than just warming. Our gluttony for energy is loading the atmosphere with pollutants and increasing the greenhouse effect. This subject is far too complicated to discuss briefly, but be assured that we can expect more frequent and more violent acts of nature, more hurricanes, [and] more heavy deluges of rain with resultant floods and mud slides—and more frequent super tornadoes.[3]

---

[1] This observation is most likely attributable to Santayana, *Reason in Common Sense*.

[2] Each of these three tragedies are discussed at length in Chapter Five of my *The Forces of Nature: Our Quest to Conquer the Planet* (Amherst, NY: Prometheus Books, 2012).

[3] Akin, *Forgotten Storm*, 8.

If Akin had published his book a few years later, no doubt he would have switched out "global warming" with "climate change." This change in vocabulary or lexicon occurred between 2000 and 2010. The new label crossed disciplinary boundaries and suggested a more nuanced political agenda than that which existed earlier in the twentieth century.

The political left is a champion for climate change legislation. In the second and third decades of the twenty-first century, Alexandria Ocasio-Cortez, a Democratic Socialist member of Congress from New York, has been a leading light among the more vocal members of the political left who seek sweeping legislative change in the name of fighting climate change. In an attempt to make a political connection to President Franklin Delano Roosevelt who promoted a domestic social program called the "New Deal," Ocasio-Cortez came up with the "Green New Deal (GND)." In an interview on National Public Radio (NPR), she claimed that "This is really about providing justice for communities. So really the heart of the GND is about social justice."[4] The fact that her proposal was labeled as insane by the author of a piece for *Money and Markets* tells us a great deal about the political waters that flow around climate change topics.[5] Whether or not her proposal is insane is not as important as the underlying goals of the initiative: more government control of the economy with an eye toward wealth redistribution. It is not clear in the least as to what impact her proposal would have on climate change, let alone improvements in the social and cultural lives of her targeted communities.

While it may seem to be beyond the scope of a book on climate change in history, it is worth also discussing that free-market capitalism, socialism, and centralized power have been tried on every inhabited continent except for Australia. There are known social and cultural factors tied to economics that affect population growth and the potential consumption of carbon containing materials, which could theoretically impact climate change, but that discussion requires a longer look. The purpose of this chapter is to provide longer, comparative looks at the interventionist proposal of the political left and the market-oriented change model of the political right. In the pages that lie ahead, we will look further into the present state of our climate with particular attention given to the role played by insolation (sunlight). The chapter shows that one of those options promises some serious social consequences and dubious environmental benefits.

Present State of the Global Climate

In Chapters 1 through 3, current climate conditions are discussed with respect to the historicity and complexity of climate change. Climate change is indeed a complex and very real topic. By the same token, what is meant by "normal climate" is also difficult to define. Do we mean average weather conditions for 5, 10, 30, or 100 years? Nevertheless, climate change is a topic that should be of concern to everyone; that concern should be tempered by an awareness that what we think is a normal climate, beyond the number of years it takes to determine normalcy, is highly dependent on many factors for which humans have little or no control. Our cognizance needs to start with solar output and solar energy absorption in the biosphere. To think that the sun always produces the same amount of energy is not supported by 400 years of data on sunspots, which serve as an indicator of solar energy output. From 1645 to 1715, for instance, the sun experienced a prolonged deficiency of sunspots that has come to be called the Maunder Minimum.[6] Another major solar minimum called the Dalton Minimum lasted from 1790 to 1830. NASA claims that minor solar minimums occur "every 11 years or so."[7] However, as is shown by the inconsistency in the number of years or duration in which a solar minimum occurred, as well as the years between them, it is best to think of solar cycles and minimums as rudimentary guesses. As was shown in early 2012, there can also be spikes in solar activity[8] that had a significant impact on the daily temperatures in North America in late winter

---

[4] Crowe, "AOC Unveils Insane Climate Plan."

[5] Ibid.

[6] Eddy, "The Maunder Minimum," 1189–1202.

[7] NASA Science, "Solar Minimum is Coming."

[8] NASA, "Geomagnetic Storm Strength Increases."

and early summer. Daily highs in the middle of March were similar to normal highs for mid-summer days in many locations.[9] Between March 14 and March 18, Chicago, Illinois, saw daily high temperatures above 80°F (26.6°C). To put that kind of temperature into perspective, Chicago only experiences one day above 80°F in a typical April. On March 18, 2012, Rochester, Minnesota, saw an overnight low of 62°F (16.7°C). That overnight low actually beat the previous record high for that date by 2°F.[10]

Since 2017, the frequency of solar sunspots has been declining. In 2019 through 2020, sunspot activity was at its lowest point since 2009.[11] Interestingly, few climate change observers in the political and media worlds have made a connection between solar output and weather, although "aberrant cold" weather has made for sensationalized news coverage. In January 2019, winter cold in North America was severe. A winter storm named Jayden blanketed much of the eastern half of the continent. In the Midwest, 1,300 flights were canceled due to wintery weather, state offices and a number of school districts in Georgia were closed, and a state of emergency was declared in Wisconsin and Alabama.[12] To give the cold weather of the winter of 2018 to 2019 a fearsome name, media outlets latched onto "polar vortex," which made the cold wave sound like a freezing tornado. The BBC reported on February 1, 2019 that the Polar Vortex had claimed 21 lives in the United States while the American Midwest was experiencing one of the worst cold snaps to hit it in decades (Figure 8.1). Ninety million Americans, according to the BBC, had experienced temperatures as low as 0°F (−17°C).[13] With the arrival of winter 2019 to 2020, weather has remained cold, especially in certain parts of the country. According to CBS Denver, the mountain city of Breckenridge, Colorado, set a new monthly record for snowfall. With eight days remaining on the February 2020 calendar, the resort town had measured 24 feet of new snow for the month.[14]

FIGURE 8.1 Polar Vortex Settles in over Chicago and a Frozen Lake Michigan.

---

[9] NASA, "Historical Heat in North American."

[10] Ibid.

[11] NOAA Space and Weather Prediction Center, "Solar Cycle Progression."

[12] Brackett and Wright, "Winter Storm Jayden."

[13] BBC, "Polar Vortex Death."

[14] CBS Denver, "Race to 300."

There is a reason why high-altitude places like Breckenridge, Vail, and Steamboat Springs, Colorado, normally experience more snowfall, albeit much less than 24 feet per month, than lower altitude locations at similar latitudes. It is because air temperatures drop by 3.5°F to 5°F per thousand feet. Incidentally, descending air masses heat up at those rates, too. This was a serious flaw in the movie *The Day After Tomorrow*.[15] In that movie, descending masses of frigid air froze objects and people as the air approached the ground. This is strictly fiction; air warms as it descends or sinks. You might note that the change in lapse rates depends on moisture content in the mass of air that is rising or sinking. Rising or plummeting moist air changes temperatures at a lower rate than dry air. Dry air, like the descending frigid air in *The Day After Tomorrow*, would actually warm up at closer to 5°F per thousand feet. This means that an airliner that is flying at 26,000 feet through air registering −60°F (−51°C) on a thermometer, which is not unusual, will be directly over people on the ground enjoying room temperature conditions at sea level. These dynamics have a tremendous impact on the weather we experience.

As air moves up and over high mountains and plateaus, the air's capacity to hold moisture declines, clouds form, and precipitation occurs. In a similar fashion, drops in air temperatures that were caused by the solar minimum created havoc across the humid southeastern United States in February 2019 and 2020.

In the Mississippi River Valley, cooler temperatures created by the solar minimum caused warm moist air fed into the region from the Gulf of Mexico to persistently drop rain for several weeks. By February 17, 2020, Tennessee had measured precipitation levels that were 400 times the normal amount for the month, and it was still 12 days short of March 1st. Weeks of rain brought "historic, unprecedented flooding" to Mississippi and Tennessee.[16] Condensation and precipitation helped to keep air temperatures from

FIGURE 8.2 Clouds Forming over Mountains.

© Smit/Shutterstock.com

---

[15] Emmerich, *Day After Tomorrow*.

[16] Rice et al., "Historic, Unprecedented Flooding Swamps Southern USA."

© LuckyTenn7/Shutterstock.com

FIGURE 8.3 Flooded Parking Lot in Knoxville, Tennessee, February 23, 2019.

dropping much lower than normal. This happened because the hydrological cycle transmits energy and heat. When water dries, it stores heat as latent energy; that is why a hot hair drier feels cool when first applied to your wet hair. As water evaporates, you begin to feel the heat from the drier. When condensation occurs, latent energy is released as heat into the atmosphere. This is why coastal places like Glasgow, Scotland, and San Francisco, California, have warmer winter temperatures than cities located at similar latitudes in the interior of their continents. Despite the patterns of solar activity and weather events, few observers openly discuss their connections. Perhaps scholars and media personalities are afraid to be labeled as climate change deniers? Being skeptical of the cause of climate change is not the same as denying the obvious reality that climate does change. Still, the negative label persists and is applied to anyone who does not blame humanity for changes in the weather.

Despite the paucity of publicly made connections between solar output and temperatures, there is a burgeoning industry in China, Europe, and North America that is harnessing battery, nuclear, wind, and solar energy to fuel everything from automobiles to watches. Solar activity or output is clearly a major factor in solar energy production through photovoltaic cells and weather. Solar irradiance is a measure of solar power or wattage per square meter; as we have already discussed, it is a function of solar angle, distance from the sun, and solar cycle.[17] I would also add atmospheric cloud cover and particulate matter that can deflect, reflect, and scatter sunlight.

---

[17] Boxwell, *Solar Electricity Handbook*, 41–42.

FIGURE 8.4 Low Albedo of Solar Panels Helps Retain More Energy.

Solar energy arrives in the earth's atmosphere mostly as shortwave radiation (see Figure 8.5). For us to experience that insolation (sunlight) as heat, it must be absorbed and then reradiated back into the atmosphere as longwave radiation or sensible heat. To more fully appreciate that process, imagine lying in the sun on a hot summer day for two hours. You will feel the heat as longwave radiation as it is emitted from your body and warms the hairs and air above your skin. If you are not used to being in the sun and have a light complexion, your intense experience with heat will arrive a few hours later as your skin emits the absorbed solar energy as sensible heat; your skin will feel hot to the touch and internally you might feel a bit of a chill as your body gives off heat. Just like our bodies, the atmosphere, earth, and water bodies also reflect, store, and emit energy. As is shown in Figure 8.5, about 19 percent of insolation is absorbed by the atmosphere and clouds, and 51 percent is absorbed at the earth's surface. That means that 30 percent of insolation is reflected back into space by the atmosphere and earth, but the majority of reflection is caused by clouds. You might recall that the reflectivity of an object is called albedo. An object with a high albedo reflects more energy back into space. On the other hand, an object with a low albedo rating absorbs a lot of shortwave radiation and is thus heated up. Heat is radiated back into the atmosphere where we feel it. You might note that the percentages given above and in Figure 8.5 are averages.

Future Climates

What do you suppose would happen to global temperatures if more cloud cover formed over the oceans for most of the warm months of the year, especially between 40°N and 40°S?[18] That potential energy would mostly be reflected back into space, the oceans would not warm, and global temperatures would drop. If skies were clear over land in that latitudinal zone, summer temperatures would not be much different than normal, but due to less energy absorbed into the tropical oceans, ocean currents like the Kuroshio or Japanese Current and Gulf Stream would not carry as much energy into zones of energy deficits (north of 40°N and south of 40°S).[19]

---

[18] The zone between 40 degrees north and south is considered to be a zone of energy surplus as more energy is absorbed in the biosphere than is emitted into space. See Mason et al., *Physical Geography*, 64.

[19] Zones of energy deficits lie north of 40°N and south of 40°S. This means that more energy is radiated into space than is absorbed at the earth's surface and atmosphere.

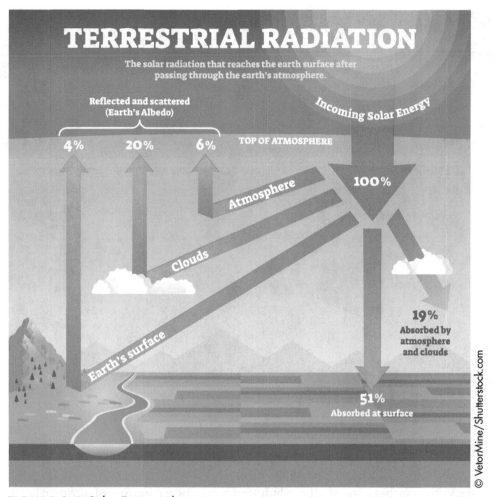

FIGURE 8.5 Solar Energy Absorption.

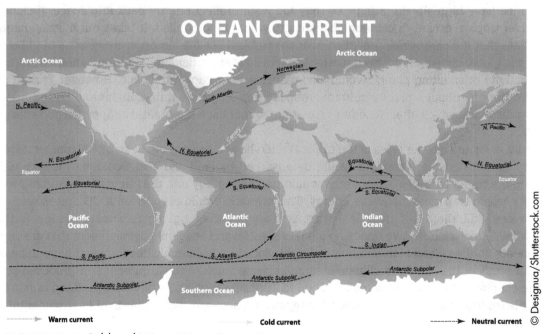

FIGURE 8.6 Cold and Warm Water Ocean Currents.

© AB Photostudio/Shutterstock.com

FIGURE 8.7 High Albedo of Snow.

In that scenario, the northern and southern latitudes would be colder. Snowfall would accumulate more and alpine glaciers would extend into lower stretches of valleys. Continental glaciers too could grow. Snow and ice have high albedo ratings, so they too would reflect energy back into space. If cloud cover remains over tropical oceans, a colder global climate would be produced. It takes but a little imagination to appreciate that if clearer skies prevailed over the tropical oceans, an opposite effect would be experienced by the global climate. It is also important to realize that particulate pollution, whether produced by volcanic activity or smelting plants, would also reflect energy back into space. However, carbon dioxide (CO_2) and other greenhouse gases would theoretically trap longwave radiation. It would be a tradeoff and anyone's guess as to which effect (cooling or heating) would have the greatest impact on societies around the world.

Ironically, clear and cleaner skies in the tropics would have a heating effect on global temperatures. This would happen because more sunlight would be absorbed by land and sea. Ocean currents would carry energy into the colder latitudes, evaporation would transfer energy into the atmosphere, and higher levels of humidity would trap more longwave energy. This scenario would occur even if CO_2 gases were significantly lowered. Still, there is a great deal of concern about human agency in climate change. In the next section, we will take a look at two opposing models of social and economic development that are advanced by proponents on the political left and political right. Social and economic development have to be considered in our study of the history of climate change because much of the argument and proposed solutions to climate change center on reducing our carbon footprint through economic change. Keep in mind that if we do not learn from the past, we are bound to repeat it. There are some aspects of environmental history

FIGURE 8.8 High Albedo of Clouds.

that we do not want to experience again while there are other past weather conditions or past geographies that we would like to experience as time unfolds. Clearly, some of the factors behind optimum weather conditions are beyond our control. How would we, for example, create weather conditions that would allow fields of clover, grape vines, and pastures to flourish in the Saudi Peninsula without irrigation? Such was the case in the seventh century.

Two Opposing Models of Social and Economic Development

The Political Left's Best Model. The model for the political left's agenda for cooling a planet made red hot by human actions is captured in the GND, which in various forms has been proposed in other countries.[20] Elements of the GND were certainly part of President Barrack Obama's environmental program. Despite wide support among Democrats, the GND was rejected in the US Senate on March 25, 2019.[21] The ideas that underpin the GND feature imposed actions that are generally aimed at perceived social and economic inequalities. According to the *Washington Post* in a piece published on February 11, 2019, the GND would provide residents with jobs boasting a guaranteed family-sustaining wage, paid vacations, family and medical leave, as well as retirement. The golden eggs do not stop with employment. Being careful not to use the term "citizen," the GND promises that the "people of the United States" would be provided with high-quality healthcare, affordable, safe, and adequate housing, economic security, and access to clean

---

[20] Harvey, "Labour's Climate Policies."

[21] Shabab and Clark, "Senate Fails to Advance Green New Deal."

air and water, affordable and healthy food, plus access to nature.[22] The proposed bill went on to promise resources for training and high-quality education for all people of the United States. When the proposed bill eventually mentioned obvious environmental topics, it promised a complete transformation of energy-based industries in the United States. In just 10 years, the GND would have the people of the United States off of carbon-based fuels. This would be aided by spurring on massive growth in clean manufacturing and by meeting 100 percent of the power demand in the United States with clean, renewable, and zero emissions energy sources. Massive investments would also be made through green energy infrastructure improvements.[23]

The intent behind the GND may well have been simple political fodder for the urban and minority poor. By pointing to stingy Republicans who are keeping them from the good life, the Democrats knew they would attract more support than they already receive from those groups. White liberals too would be compelled to support the more compassionate Democrat Party. From a political standpoint, the Republicans seem to be asleep at the wheel. They have said little about climate change, so they have been easy targets for making negative labels stick to them. Their opponents call them climate deniers. By extension, if one is a climate denier, he or she must be a science denier. By further extension, if one is a science denier, he must be a knuckle dragging Neanderthal and so on. In reality, many of the environmental ideas that are contained in the GND are already gaining ground in the market place. Perhaps that is why Republicans are relatively quiet on the subject.

Market Forces and Energy Efficiency Model of the Political Right

There is wisdom in old sayings. An old English proverb states "necessity is the mother of invention." This saying is consistent with the human ecology paradigm in geography that was advanced by University of Chicago Professor Harlan Barrows (1877–1960). In a 1922 presidential address before the Association of American Geographers (AAG), Barrows argued that people encounter difficult situations, which are often products of their environments; they make adjustments or adaptations in their cultures, so they can survive and thrive in the face of what nature throws at them. Their adaptations often feature inventions.[24]

As the cost of fossil fuels has climbed, the profit potential for more efficient energy sources has likewise increased. With the state American-based company Tesla leading the way, the geographic range and practicality of electric vehicles (EVs) is exploding.[25] Not only are these vehicles efficient, they are among the fastest on the road.[26] In 2019, there were 3.2 million electric cars sold in the international marketplace. It is expected that if present trends continue, just under 27 million vehicles will be sold in the year 2030. In countries like China and Mexico where air pollution is a major health risk, there is increasing interest in electric cars, trucks, bicycles, and motorcycles. China is not waiting on foreign companies to meet the demand for EVs. Among the dozen or so Chinese EV manufacturers is BYD Company LTD. The company's stocks are doing well in the American stock exchanges.[27]

To recharge an EV's battery, though, one needs a reliable source of rapidly dispensed electricity. In 2018, the United States had 20,021 charging stations and 57,187 charging outlets.[28] While most of those outlets and stations were in California, states like Colorado, which has a high level of environmental awareness and demand for EVs like Tesla cars, are experiencing growth in an industry that provides charging stations and outlets (see Figures 8.9 and 8.10).

---

[22] Rizzo, "What's Actually in the Green New Deal."

[23] Ibid.

[24] Barrows, "Geography as Human Ecology," 1–14.

[25] Markets and Markets, "Electric Vehicle Market."

[26] Kane, "Tesla Cars Dominate."

[27] Yahoo Finance, "BYD Company Limited."

[28] Wagner, "Number of Public Electric Vehicle Charging Stations."

FIGURE 8.9 Electric Vehicle Charging Station.

FIGURE 8.10 An Electric Vehicle (EV) Hooked to Charging Outlet.

Tesla, no doubt, has led the way in EV innovation, production, and sales. The EV industry is experiencing what economists call an economic profit. Industries are most likely to expand during periods of time in which economic profits are made. An economic profit occurs when a firm's total revenue exceeds total explicit and implicit costs. For such a condition to exist, there must be high demand for the product. If the demand exceeds the supply of the product and those firms operating in the industry are experiencing economic profits, growth in the industry is likely to occur. While Toyota and many other auto manufacturers have produced hybrid vehicles for well over a decade, production of all electric cars is expanding.

As more and more firms operate in the expanding EV market to meet consumer demand, prices will come down. In 2020, Audi, Chevrolet, and BMW are enjoying sales of their E-Tron, Bolt, and i3 respectively. Ford Motor Company is working to compete with Tesla's Model Y and Roadster by revealing plans to begin selling its sporty Mustang in a Mach-E GT, all electric model in 2021 (see Figure 8.11). One of the major criticisms of EVs has been their lack of range and the time it takes to recharge them. Ford's Mach-E GT is designed to travel 300 miles on a single charge, which is similar to Tesla's Model Y. That kind of range makes EVs capable of much more than just urban commuting.[29]

To make this model appealing to a younger-aged market, the Mach-E GT is engineered to go from 0 to 60 MPH in the mid three-second range. This is a quicker speed than the current gas-powered Mustang GT, the company's legendary muscle car.[30]

The EV market is not just about automobiles. Tesla is producing a class 8 semi-truck with a range of 300 to 500 miles, and the legendary American motorcycle company Harley Davidson is producing an electric motorcycle handily named Livewire (see Figure 8.12).

FIGURE 8.11 Ford's Mach-E GT Mustang at a Chicago Auto Show in 2020.

© Dawid S Swierczek/Shutterstock.com

---

[29] Loveday, "Pre-Production Ford Mustang Mach-E Rolls off."

[30] Florea, "Mustang Muscle Car."

FIGURE 8.12 Harley Davidson's All Electric Livewire.

FIGURE 8.13 Kansas Windfarm.

As these innovations show, without the heavy hand of government, the marketplace is already engaged in delivering technology that can address much of the environmental objectives of the GND. Energy innovations are not just focused on solar and battery powered generation technology. In places like the American Great Plains, a substantial portion of electric power is generated by wind turbines (see Figure 8.13).

In the United Kingdom of Great Britain and Northern Ireland, major shifts have taken place in the manner in which its electric power is generated. In 2012, for instance, 42 percent of its electricity was generated by coal and oil, and 20 percent was generated by nuclear plants. In 2017, oil and coal only accounted for seven percent of its power generation while nuclear power increased to 24 percent. Over the same time period, solar and wind-power production in the United Kingdom more than doubled from 10 to 23 percent. Natural gas, which burns more efficiently than other fossil fuels, climbed from 27 percent to 45 percent of the country's source for electric power generation.[31]

The economic behemoths in the Pacific are China, Korea, and Japan (see Figure 8.14). The Japanese experienced a nuclear power plant accident at Fukushima in 2011. Between September 2013 and August 2015, Japan temporarily suspended its nuclear power program to conduct a full safety inspection. By November 2018, nine of the country's 34 operable nuclear power plants were back in operation. Five of those were reactivated in 2018.

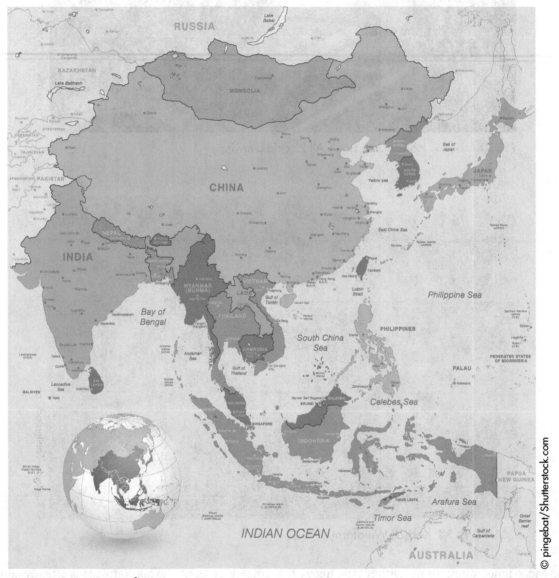

FIGURE 8.14 Map of Asia.

© pingebat/Shutterstock.com

---

[31] US Energy Information Administration, "Coal Power Generation Declines."

In 2017, natural gas was Japan's primary energy source for generating electricity at 37 percent. Natural gas was followed closely by imported coal, which accounted for one-third of the country's electric power generation. Japan has a goal to increase the percentage of its electric power generation through nuclear technology to 22 percent by 2030.[32] To the economically minded Japanese, nuclear power offers the greatest amount of energy at the lowest cost. Accidents aside, nuclear power represents a near inexhaustible supply of energy with limited environmental impacts.

Human Population Growth and Economic Development

As the human population continues to grow on a global scale, we can expect there to be an increase in the demand for power and consumable products, which is theoretically associated with climate change. Paradoxically, economic development is associated with resource exploitation, but it is also shown to be an effective means to reign in population growth. As I discussed in *The Forces of Nature: Our Quest to Conquer the Planet*, the growth in human population is not uniform across the world's landscape. Indeed, there are serious differences in population growth rates in various countries. As a general rule, 30°N divides the world into low-growth countries and high-growth countries. There is a significant, inverse correlation between levels of economic development and population growth; therefore, we can state with confidence that as economic development increases, population growth rates decline. A number of developed countries that are situated north of 30°N now have zero and even negative natural increase rates. This phenomenon is rather easy to explain: in more developed countries, women who possess the same cognitive potential as men, are encouraged to contribute their talents in the workplace. As a result, there is a corresponding de-emphasis on childbearing as a major function of being a woman. With fewer women having babies, natural increase falls.

SUMMARY

As the range of topics in this chapter shows, climate change is a highly complex issue that requires a grounding in the natural sciences as well as a keen appreciation, if not understanding, of the social and economic factors that affect demographic transitions, resource exploitation, and atmospheric pollutants.

In this book, we have looked at the historicity and complexity of climate change. In doing so, we introduced a compelling new look into past geographies that were impacted by climates and climate change. This nuanced approach was accomplished by applying the framework captured in John K. Wright's geotheological lexicon. We found that few studies have considered the practices and narratives of ancient belief systems and religious writings to help us gauge the degree to which climate has changed. We have seen that biblical texts like the Book of Job, the Pentateuch, the history books, and the post Babylonian exile books provide narratives that certainly seem to confirm the data and descriptions of past climates found in a number of paleoclimatological studies; moreover, we found that New Testament writings seem to confirm a climate epoch that historians call the Roman Warm Period. Moving forward in time, we see that the narratives in the Quran suggest a wetter climate in the Saudi Peninsula that provided sufficient rainfall to support fields of clover, grape vines, and pastures on which cattle could graze.

Perhaps most importantly, we showed how the sun is a major factor in the heating and cooling of our atmosphere and oceans. Solar activity varies from time to time and may not ultimately match the predicted schedule that scientists have created in the anticipation of the recurrence of solar sunspots, solar minimums, and solar maximums. The earth's tilt, too, has changed from time to time. Volcanic activity releases greenhouse gases and particulate matter that can heat and cool the atmosphere.

We humans no doubt have some effect on climate because we know that certain gases like methane and CO_2 retain longwave radiation. Water vapor, or humidity, is also a greenhouse gas. We also described

---

[32] US Energy Information Administration, "Japan has Restarted Five Nuclear Power Reactors."

how nature stores many of those greenhouse gases, which would push us toward yet another cold epoch and maybe even an ice age.

Since humans can affect the greenhouse storage process, we have to consider which route toward a sustainable future is best for us to take. This chapter shows that the marketplace is moving us toward a more sustainable future that would not only provide us with more efficient energy systems, but it would encourage the use of the many talents that women have to offer in the marketplace of ideas and products. While motherhood is a precious role to play, economic incentives will draw more and more of the world's female population into the workplace. If the pattern in the developed world remains constant and its economic and social changes spread throughout the developing world, which is certainly the case, the future will lead to reduced population growth rates and resource exploitation through the innovation of more cost effective and energy efficient technologies. This desirable situation is being achieved without the strong arm of centralized authority, which enacting bills like the GND would require.

Bibliography

AGRI-LMI Labour Market Information. "Canada's Agriculture Sector Labour Market Forecast to 2025." https://cahrc-ccrha.ca/sites/default/files/files/Labour-Employment/factsheet_NAT_E.pdf.

Akin, Wallace. *Forgotten Storm: The Great Tri-State Tornado of 1825*. Guilford, CT: The Lyons Press, 2002.

Allen, Leslie. *The Books of Joel, Obadiah, Noah & Micah: New International Commentary on the Old Testament*. 2nd ed. Grand Rapids, MI: Eerdmans, 1976.

Atkinson, Kenneth, "Noble Deaths at Gamla and Masada? A critical Assessment of Josephus' Accounts of Jewish Resistance in Light of Archaeological Discoveries" in *Making History: Josephus and Historical Method*, ed. Zuleika Rogers, 349–371. Leiden, The Netherlands: Brill Publishers, 2007.

Austen, Jane. *Pride and Prejudice*. Whitehall, England: Thomas Edgerton, 1813.

———, *Sense and Sensibility*. Whitehall, England: Thomas Edgerton, 1811.

Bar-Mathews, M., and A. Ayalon. "Mid-Holocene Climate Variations Revealed in High-Resolution Speleothem Records from Soreq Cave, Israel and Their Correlations with Cultural Changes." *The Holocene* 21, no. 1 (2011): 168.

Barrows, Harlan J. "Geography as Human Ecology." *Annals of the Association of American Geographers* 13, no. 1 (1923): 1–14.

BBC. "Polar Vortex Death Toll Rises to 21 as US Cold Snap Continues." https://www.bbc.com/news/world-us-canada-47088684.

Black, Henry Campbell. *Black's Law Dictionary*. 6th Ed. Saint Paul, MN: West Publishing, 1996.

Boxwell, Michael. *Solar Electricity Handbook: A Simple, Practical Guide to Solar Energy; How to Design and Install Photovoltaic Solar Electric Systems, 2012 Edition*. Warwickshire, UK: Greenstream, 2012.

Brackett, Ron, and Pam Wright. "Winter Storm Jayden: More than 1,300 Flights Cancelled, State Offices and Schools Closed in Midwest, South; 3 Dead," *The Weather Channel*. https://weather.com/news/news/2019-01-28-winter-storm-jayden-impacts-preps.

Brecht, Martin. *Martin Luther: His Road to Reformation 1483–1521*. Minneapolis, MN: Fortress Press, 1993.

Brindle, W. A. C. J. Diemer, E. G. Dobson, J. Falwell, P. R. Fink, J. A. Freerksen, E. A. Hindson, et al., eds. *The King James Study Bible*. Nashville, TN: Thomas Nelson Publishers, 1988.

Callaway, Ewen. "Genetic Adam and Eve Did Not Too Far Apart in Time: Studies Re-date 'Y-Chromosome Adam' and Mitochondrial Eve,'" *Nature*: 1. August 6, 2013. https://www.nature.com/news/genetic-adam-and-eve-did-not-live-too-far-apart-in-time-1.13478#/ref-link-2.

Carty, Daniel. "Metrodome Roof Collapse Caught on Tape," *CBS Sports*, December 13, 2010. http://www.cbsnews.com/8301-31751_162-20025435-10391697.html.

CBS Denver. "Race to 300: Steamboat Springs and Breckenridge Getting Record Snow." https://denver.cbslocal.com/2020/02/19/steamboat-springs-breckenridge-record-snow/.

Clark, Audrey N. *The Penguin Dictionary of Geography*. 3rd ed. London: Penguin Books, 2003.

Climate-Data.org. "Climate Saudi Arabia: Weather by Month for Saudi Arabia." https://en.climate-data.org/asia/saudi-arabia-29/.

Crowe, J. T. "AOC Unveils Insane Climate Plan Backed by Social Justice Warriors, 2020 Dem Hopefuls." *Money and Markets*, February 7, 2019. https://moneyandmarkets.com/ocasio-cortez-green-new-deal-2/.

Crowley, Thomas J, and Thomas S. Lowery. "How Warm was the Medieval Warm Period," *AMBIO: A Journal of the Human Environment* 29, no. 1 (2000): 51–54.

Daley, Jason. "It's the Bicentennial of 'Silent Night'" *Smithsonian Magazine*, December 17, 2018. https://www.smithsonian-mag.com/smart-news/silent-night-celebrates-its-bicentennial-180971044/.

Darby, H. C. *Relations of History and Geography: Studies in England, France, and the United States*. Exeter, UK: Exeter University Press, 2002.

Diaz, Henry F., and Hughes, M. *The Medieval Warm Period*. Boston: Kluwer Academic Publishers, 1994.

Domosh, Mona, Roderick P. Neumann, and Patricia L. Price. *Contemporary Human Geography: Culture, Globalization, and Landscape*. New York: W. H. Freeman, 2015.

Douglas, J. D., F. F. Bruce, J. I. Packer, N. Hillyer, D. Guthrie, A. R. Millard, and D. J. Wiseman. *The New Bible Dictionary*. 2nd ed. Wheaton, IL: Tyndale House, 1991.

Drake. B. L. "The Influence of Climate Change in the Late Bronze Age Collapse and the Greek Dark Ages." *Journal of Archaeological Science* 39, no. 6 (June 2012): 1862.

Dunn, Rob. "Mistletoe: The Evolution of a Christmas Tradition." *Smithsonian Magazine*, December 21, 2011. https://www.smithsonianmag.com/science-nature/mistletoe-the-evolution-of-a-christmas-tradition-10814188/.

Durkheim, Émile. *The Elementary Forms of the Religious Life*. London: George Allen & Unwin, 1912.

East, W. Gordon. *Geography behind History*. New York: W. W. Norton, 1999.

Eddy, J. A. "The Maunder Minimum." *Science* 192, no. 4245 (June 1976): 1189–1202.

Emmerich, Robert. *The Day after Tomorrow*. Los Angeles: Twentieth Century Fox, 2004.

ESPN. "Fierce Storm Sacks Vikings-Eagles Game," https://www.espn.com/nfl/news/story?id=5956740.

Filmer, W. E. "Chronology of the Reign of Herod the Great," *Journal of Theological Studies* XVII, no. 2 (1966): 283–298.

Fleming, K. "Cultural Landscape: A Theoretical Perspective." *Proceedings of the Society for California Archaeology* 11 (1997): 112–117. https://scahome.org/publications/proceedings/Proceedings.11Fleming.pdf.

Florea, Ciprian. "Is the Mustang Muscle Car Becoming Irrelevant." *TopSpeed*, November 2019. https://www.topspeed.com/cars/car-news/the-2021-ford-mustang-mach-e-gt-will-be-quicker-than-the-mustang-gt-ar187064.html.

Fox, Michael V. *The JPS Bible Commentary: Ecclesiastes*. Philadelphia: Jewish Publication Society, 2004.

Garfinkel, Harold. "The Origins of the Term Ethnomethodology." In *Ethnomethodology: Selected Readings,* edited by R. Turner, 15–18. London: Penguin Education, 1974.

Goddard Institute for Space Studies, "2009: Second Warmest Year on Record; End of Warmest Decade," *NASA*, January 21, 2010. http://www.giss.nasa.gov/research/news/20100121/.Graham, Steve. "Milutin Milankovitch (1879–1958)." *NASA Earth Observatory*, (Originally published March 24, 2000). https://earthobservatory.nasa.gov/features/Milankovitch/milankovitch_2.php.

Gribben, J., and H. H. Lamb. "Climate Change in Historical Times." In *Climate Change*, edited by J. Gribben, 68–82. Cambridge: Cambridge University Press, 1978.

The Guardian, "The Ice Age Cometh." 12 June 1962.

Hansen, James, Makiko Sato, and Redo Ruedy, *Public Perception of Climate Change and the New Climate Dice*. New York: NASA Goddard Institute for Space Studies and Columbia University Earth Institute, 2012.

Harvey, Fiona. "Labour's Climate Policies: What are They and What Do They Mean?" *The Guardian*, September 24, 2019. https://www.theguardian.com/politics/2019/sep/24/labours-climate-policies-what-are-they-and-what-do-they-mean.

Hayes, J. D., John Imbrie, and N. J. Shackleton. "Variations in the Earth's Orbit: Pacemaker of the Ice Ages." *Science* 194 (December 1976): 1121–1132.

Hendy, Erica J., Michael K. Gagan, Chantal A. Alibert, Malcolm T. McCulloch, Janice M. Lough, and Peter J. Isdale. "Abrupt Decrease in Tropical Pacific Sea Surface Salinity at End of Little Ice Age." *Science* 295, no. 5559 (2002): 1511–1514.

History Channel. "First Groundhog Day." https://www.history.com/this-day-in-history/first-groundhog-day.

History.com. "Groundhog Day: History and Facts," February 2, 2012. https://www.history.com/news/groundhog-day-history-and-facts.

Hoyt, Richard. "When to Plant Clover Seeds." *Hunker*, June 22, 2018. https://www.hunker.com/13426433/when-to-plant-clover-seeds.

Hughes, Malcolm K., and Henry F. Diaz, "Was there a Medieval Warm Period, and if so, Where and When." *Climate Change* 26, no. 2–3 (1994): 109–142.

Hulme, Mike. "The Conquering of Climate: Discourses of Fear and Their Dissolution." *The Geographical Journal* 174, no. 1 (March 2008): 5–16.

Independent. "2010 Showed Coldest December on Record." https://www.independent.co.uk/news/uk/home-news/2010-showed-coldest-december-on-record-2176729.html.

Institute for Creation Research. "Discover and Defend." http://www.icr.org/books/defenders/2603/band.

Kane, Mark. "Tesla Cars Dominate 0-60 MPH, But I Pace Time Might Surprise You." *Motor1.com*. https://www.motor1.com/news/315872/tesla-electric-0-60-mph-ipace/.

Keigwin, Lloyd D. "The Little Ice Age and Medieval Warm Period in the Saragossa Sea." *Science* 274, no. 5292 (1996): 1503–1508.

Kennedy, John F., "Remarks at the America's Cup Dinner Given by the Australian Ambassador, September 14, 1962." *Presidential Library and Museum*. https://www.jfklibrary.org/archives/other-resources/john-f-kennedy-speeches/americas-cup-dinner-19620914.

Kim, P. J. "Letters of John." In *Westminster Dictionary of the New Testament and Early Christian Literature*, edited by D. Aune, 249–251. Westminster: John Knox Press, 2003.

Knecht, Friedrich Justus. *A Practical Commentary on Holy Scripture*. London: B. Herder, 1910.

Kugler, R. A., and P. J. Hartin, *An Introduction to the Bible*. Grand Rapids, MI: William B. Eerdmans Publishing, 2009.

Lamb, H. H. *The Changing Climate: Selected Papers*. New York: Routledge, 2011.

Lee, Richard D. ed. *The American Patriots Bible: The Word of God and the Shaping of America*. Nashville, TN: Thomas Nelson, 2009.

Lewis, Tanya. "Groundhog Day: How Accurate is Punxsutawney Phil Anyway?" *The Christian Science Monitor*, February 2, 2015.

Loveday, Eric. "Pre-Production Ford Mustang Mach-E Rolls Off Assembly Line in Mexico." *Inside EVs*, February 2020. https://www.bing.com/search?q=ford+mustang+mach+e&qs=LC&pq=ford+mustan&sk=EP1&sc=8-11&cvid=0D5A70C1D-DA141769F411169D5F088DE&FORM=QBRE&sp=2.

Maclean, Norman. *A River Runs Through it and Other Stories*. Chicago: University of Chicago Press, 1976.

Mann, Michael E. "Medieval Climatic Optimum." In *Encyclopedia of Global Environmental Change Volume 1*, edited by Michael C. McCracken and John S. Perry. Chichester, England: John Wiley & Sons, 2002.

Markets and Markets. "Electric Vehicle Market." https://www.marketsandmarkets.com/Market-Reports/electric-vehicle-market-209371461.html.

Mason, Joseph A., James E. Burt, Peter O. Muller, and H. J. de BLIJ. *Physical Geography: The Global Environment*. 5th ed. Oxford: Oxford University Press, 2016.

McCormick, Michael, Ulf Büntgen, and Mark A. Cane. "Climate Change During and After the Roman Empire: Reconstructing the Past from Scientific and Historical Evidence," *Journal of Interdisciplinary History* XI III, no. 2 (Autumn 2012): 169–220.

Met Office, "Snow and Low Temperatures, December 2010." https://www.metoffice.gov.uk/binaries/content/assets/metoffice-govuk/pdf/weather/learn-about/uk-past-events/interesting/2010/snow-and-low-temperatures-december-2010---met-office.pdf.

Migowski, Claudia, Mordechai Stein, Sushma Prasad, Jörg F. W. Negendank, and Amotz Agnon. "Holocene Variability and Cultural Evolution in the Near East from the Dead Sea Sedimentary Record." *Quaternary Research* 66, no. 3 (November 2006): 421–431.

Mullan, David George. *Scottish Puritanism, 1590–1638*. Oxford: Oxford University Press, 2000.

NASA. "Arctic and Antarctic Sea Ice: How are they Different?" https://climate.nasa.gov/blog/2861/arctic-and-antarctic-sea-ice-how-are-they-different/.

———, "Geomagnetic Storm Strength Increases." https://www.nasa.gov/mission_pages/sunearth/news/News030712-X5-4.html.

———, "Historical Heat in North American Turns Winter to Summer." https://earthobservatory.nasa.gov/images/77465/historic-heat-in-north-america-turns-winter-to-summer.

———, "Sunspot and Flare, March 2012." *Earth Observatory*. https://earthobservatory.nasa.gov/images/77321/sunspot-and-flare-march-2012.

———, "What's the Difference between Weather and Climate." February 1, 2005. https://www.nasa.gov/mission_pages/noaa-n/climate/climate_weather.html.

NASA Science. "Solar Minimum is Coming." https://science.nasa.gov/science-news/news-articles/solar-minimum-is-coming.

National Weather Service. "Snow Records for Pittsburg, Pennsylvania." https://www.weather.gov/pbz/tsnow#LARGEST.

NBC New York. "2016 Blizzard was NYC's Biggest Snowstorm on Record, NOAA Report Finds." https://www.nbcnewyork.com/news/local/nyc-new-york-city-blizzard-biggest-ever-january-23-2016/831660/.

NOAA. "National Snow and Ice—January 2011," https://www.ncdc.noaa.gov/sotc/snow/201101.

NOAA Space and Weather Prediction Center. "Solar Cycle Progression." February 2020. https://www.swpc.noaa.gov/products/solar-cycle-progression.

Owen, Oliver S. *Natural Resource Conservation: An Ecological Approach*. 3rd ed. New York: Macmillan Publishing, 1980.

Park, Chris. *Sacred Worlds: An Introduction to Geography and Religion*. Abingdon, UK: Routledge, 1994.

Patlakas, Platon, Christos Stathopoulos, Helena Flocas, Christina Kalogeri, and George Kallos. "Regional Climatic Features of the Arabian Peninsula." *Atmosphere* 10, no. 4 (2019): 220.

Penna, Anthony N. "Climate Change." In *Berkshire Encyclopedia of World History*. 2nd ed., edited by William H. McNeil, Jerry H. Bentley, David, Christian, Ralph C. Croizier, J. R. McNeil, Heidi Roupp, and Judith P. Zinsser. Great Barrington, MA: Berkshire Publishing Group, 2011.

Perry, Charles A., and Kenneth J. "Geophysical, Archaeological, and Historical Evidence Support a Solar-output Model for Climate Change." *Proceedings of the National Academy of Sciences of the United States of America* 97, no. 23 (2000): 12433–12438.

Philo, Chris. "History, Geography, and the Still Greater Mystery of Historical Geography." In *Human Geography: Society, Space, and Social Science*, edited by Derek Gregory, Ron Martin, and Graham Smith. Minneapolis, MN: University of Minnesota Press, 1994.

Proyas, Alex. *Knowing*. Los Angeles: Summit Entertainment, 2009.

Radmacher, Earl D., Ronald B. Allen, and H. Wayne House, eds. *New King James Version*. 2nd ed. Nashville, TN: Thomas Nelson, 2007.

Ranken Energy Corporation. "Products Made from Petroleum." https://www.ranken-energy.com/index.php/products-made-from-petroleum/.

Ravenstein, E. G. *The Laws of Migration*. Santa Barbara, CA: John Corbett, Center for Spatially Integrated Social Science, (original publication 1885).

Redford, Robert Director. *A River Runs Through It*. Los Angeles: Columbia Pictures, 1992.

Reynolds, Kevin. *Water World*. Los Angeles: Universal Pictures. 1995.

Rice, Doyle, Luke Ramseth, and Wilton Jackson. "Historic, Unprecedented Flooding Swamps Southern USA; Mississippi and Tennessee Hardest Hit." *USA Today*, https://www.usatoday.com/story/news/nation/2020/02/17/mississippi-flooding-swamps-southern-us/4784911002/.

Rizzo, Salvador. "What's Actually in the Green New Deal from Democrats," *Washington Post*, February 11, 2019. https://www.washingtonpost.com/politics/2019/02/11/whats-actually-green-new-deal-democrats/.

Rogerson, J. W. "The History of Tradition: Old Testament and Apocrypha." In *Eerdmans Commentary on the Bible*, edited by D G. Dunn and J. W. Rogerson. Grand Rapids, MI: Eerdmans, 2003.

Santayana, George. *In his Reason in Common Sense*. New York: Dover Publishing, 1905.

Scheihing, Mark H., and Hermann W. Pfefferkorn. "Taphonomy of Land Plants in the Orinoco Delta: A Model for the Incorporation of Plant Parts in Clastic Sediments of Late Carboniferous Age of Euramerica." *Review of Palaeobotany and Palynology* 41, no. 3–4 (1984): 205–240.

Schwehdi, M. H. "Thunderstorm Frequency and Distribution in Saudi Arabia." *Journal of Geophysics and Engineering* 2, no.1 (2005): 252.

Shabab, Rebecca, and Dartunorro Clark. "Senate Fails to Advance Green New Deal as Democrats Protest McConnell's Sham Vote." *NBC News.* https://www.nbcnews.com/politics/congress/senate-fails-advance-green-new-deal-democrats-protest-mcconnell-sham-n987506.

Sheppard, Si. *The Jewish Revolt AD 66–74.* Oxford: Osprey Publishing, 2013.

Siklosy, Z., A. Demeny, T.W. Vennemann, J. Kramers, S. E. Lauritzen, and S. Leel-Ossy, "Middle Bronze Age Climate Change Recorded in Hungarian Stalagmite: Triggering by Volcanic Activity?" *Geophysical Research Abstracts*, 9 (2007): http://www.cosis.net/abstracts/EGU2007/00777/EGU2007-J-00777.pdf.

Stein, Mordechai, Adi Torfstein, Ittai Gavrieli, and Yoseph Yechieli. "Abrupt Aridities and Salt Deposition in the Post-Glacial Dead Sea and Their North Atlantic Connection." *Quaternary Science Reviews* 29 (2010): 567–575.

Steinmann, Andrew. *From Abraham to Paul: A Biblical Chronology.* St. Louis, MO: Concordia Publishing House, 2011.

Stothers, Richard B. "The Great Tambora Eruption in 1815 and its Aftermath." *Science* 224, no. 4654 (1984): 1191–1198.

Stromberg, Joseph. "The Age of Humans: Living in the Anthropocene." *Smithsonian Magazine*, January 2013. https://www.smithsonianmag.com/science-nature/what-is-the-anthropocene-and-are-we-in-it-164801414/.

Swiss National Science Foundation. "Sun's Impact on Climate Change Quantified for First Time." https://phys.org/news/2017-03-sun-impact-climate-quantified.html.

Symonds, R. B, eds. Carroll, M. R. and Holloway, J. R. "Volatiles in Magmas." *Reviews in Mineralogy* 30 (1994): 1–66.

Tourist Israel: The Guide. "Mount Hermon Ski Resort." https://www.touristisrael.com/mount-hermon-ski-resort/2446/.

Trotta, Daniel. "New York Hard hit as Winter Storm Slams Northeast," *Reuters: Edition US*, December 27, 2010, http://www.reuters.com/article/idUSTRE6BP1EW20101227.

Tuan, Yi-Fu. *Morality and Imagination: Paradoxes of Progress.* Madison, WI: University of Wisconsin, 1989.

United States Geological Survey. "Cataclysmic 1991 Eruption of Mount Pinatubo, Philippines." http://pubs.usgs.gov/fs/1997/fs113-97/.

US Climate Data. https://www.usclimatedata.com/climate/denver/colorado/united-states/usco0105.

US Energy Information Administration. "Coal Power Generation Declines in United Kingdom as Natural Gas, Renewables Grow." *Today in Energy* (April 2018), https://www.eia.gov/todayinenergy/detail.php?id=35912.

———, "Japan has Restarted Five Nuclear Power Reactors in 2018." *Today in Energy*, November 2018. https://www.eia.gov/todayinenergy/detail.php?id=37633.

Vann, Barry Aron. *The Forces of Nature: Our Quest to Conquer the Planet.* Amherst, NY: Prometheus Books, 2012.

———. *In Search of Ulster-Scots Land: The Birth and Geotheological Imagings of a Transatlantic People, 1603–1703.* Columbia: University of South Carolina Press, 2008.

———. *Rediscovering the South's Celtic Heritage.* Johnson City, TN: Overmountain Press, 2004.

Villalba, Ricardo. "Tree Ring and Glacial Evidence for the Medieval Warm Epoch and the Little Ice Age in Southern South America." *Climate Change* 26, no. 2–3 (1994): 183–197.

Wagner, I. "Number of Public Electric Vehicle Charging Stations and Charging Outlets in the U.S. as of December 2018." *Statista.* https://www.statista.com/statistics/416750/number-of-electric-vehicle-charging-stations-outlets-united-states/.

Wang, Ting, Donna Surge, and Karen Jo Walker. "Seasonal Climate Change Across the Roman Warm Period/Vandal Minimum Transition Using Isotope Sclerochronology in Archeological Shells and Otoliths, Southwest Florida, USA." *Quaternary International* 308–309 (October 2013): 230.

Ward, D. "Climate of the Holocene." ATMO 336. University of Arizona. http://www.atmo.arizona.edu/students/courselinks/fall12/atmo336/lectures/sec5/holocene.html.

Weather Channel. "Data for Amman, Jordan," https://weather.com/weather/monthly/l/JOXX0002:1:JO.

Weather.com. "The Most Extreme Wind Speed Ever Recorded on Earth." https://weather.com/storms/severe/news/2018-04-10-most-extreme-winds-earth-surface.

Weather Online. "Iraq," https://www.weatheronline.co.uk/reports/climate/Iraq.htm.

Weaver, Dorothy Jean. "Inheriting the Earth: Towards a Geotheological Narrative of Matthew's Narrative." *The Journal of Inductive Biblical Studies* 2, no. 1 (2015): 6–29.

Werner, Johannes, Jianghao Wang, Juan José Gómez Navarro, Nathan Steiger, and Raphael Neukom. "No Evidence for a Globally Coherent Warm and Cold Periods over the Pre-Industrial Common Era." *Nature* 571, no. 7766 (2019): 550–554.

Wolf, A. T. *Hydropolitics along the Jordan River: Scarce Water and Its Impact on the Arab-Israeli Conflict.* New York: United Nations University, 1995.

World Atlas. "Does it Snow in Jerusalem? Does it snow in Israel?" https://www.worldatlas.com/articles/does-it-snow-in-jerusalem-does-it-snow-in-israel.html.

World Meteorological Association (WMO), cited in *The World Almanac and Book of Facts 2015.* New York: World Almanac Books, 2015.

Wright, John K. *Human Nature in Geography: Fourteen Papers, 1925–1965.* Cambridge: Harvard University Press, 1966.

———, "Terrae Incognitae: The Place of Imagination in Geography," *Annals of the Association of American Geographers* 37 (1947): 1–45.

Yahoo Finance. "BYD Company Limited." https://finance.yahoo.com/quote/BYDDF/.

Ynetnews. "Mini Tornado Sweeps through Western Galilee," last modified May 2006, http://www.ynetnews.com/articles/0,7340,L-3236346,00.html.

Zakai, Avihu. *Exile and Kingdom: History and Apocalypse in the Puritan Migration to America.* Cambridge: Cambridge University Press, 2002.

Zias, Joe. "Human Skeletal Remains from the Southern Cave at Masada and the Question of Ethnicity." In *The Dead Sea Scrolls Fifty Years after Their Discovery*, edited by L. Schiffman, J. VanderKam, and M. Emanuel. Jerusalem: Israel Exploration Society, 2000: 732–738.

Index